FM 21-45

US WAR DEPARTMENT BASIC FIELD MANUAL

PROTECTIVE MEASURES, INDIVIDUALS AND SMALL UNITS

1942

WORLD WAR II

CIVILIAN REFERENCE EDITION

UNABRIDGED HISTORIC ARMY HANDBOOK REGARDING PERSONAL CAMOUFLAGE, CONCEALMENT, AND SOLDIER DEFENSIVE MEASURES

U.S. WAR DEPARTMENT

Doublebit Press

www.doublebitpress.com | Cherry, IL, USA

*Original content under the public domain; unrestricted for civilian distribution.
Originally published in 1942 by the U.S. War Department.*

A Part of the Military Outdoors Skills Series

*Doublebit Press Civilian Reference Edition ISBNs
Hardcover: 978-1643892146
Paperback: 978-1643892153*

WARNING: Some of the material in this book may be outdated by modern safety standards. This antique text may contain outdated, unsafe, or now illegal activities, projects, or mechanical, electrical, chemical, or medical practices and operations. Any use of this book for purposes other than historic study may result in unsafe and hazardous conditions and individuals act at their own risk and are responsible for their own safety. Doublebit Press, its authors, or its agents assume no liability for any injury, harm, or damages to persons or property arising either directly or indirectly from any content contained in this text or the activities performed by readers. Remember to be safe with any activity or work you do, use good judgement by following proper health and safety protocols, and be properly trained. In addition, because this book was from a past time and is presented in an unabridged form, the contents may be culturally or racially insensitive. Such content does not represent the opinions or positions of the publisher and are presented for historical posterity and accuracy to the original text.

DISCLAIMER: Doublebit Press has not tested or analyzed the methods, materials, and practices appearing in this public domain text and provides no warranty to the accuracy and reliability of the content. NOT CURRENT MILITARY DOCTRINE: no military operations or methods described within are endorsed or suggested by the publisher and this text is not intended to be used in any actual operations or programs. This text is published for entertainment, historical study, and personal literary enrichment purposes only and should only be used for such. This text is provided only as a reprinted facsimile from the unedited public domain original as first published and authored by the original author and publisher. The publisher assumes no liability for any injury, harm, or damages to persons or property arising either directly or indirectly from any information contained in this public domain book or activities performed by readers.

FM 21-45

BASIC FIELD MANUAL

PROTECTIVE MEASURES, INDIVIDUALS AND SMALL UNITS

UNITED STATES
GOVERNMENT PRINTING OFFICE
WASHINGTON : 1942

WAR DEPARTMENT,
WASHINGTON, MARCH 10, 1942.

FM 21–45, Protective Measures, Individuals and Small Units, is published for the information and guidance of all concerned.

[A. G. 062.11 (12–19–41).]

BY ORDER OF THE SECRETARY OF WAR:

G. C. MARSHALL,
Chief of Staff.

OFFICIAL:

J. A. ULIO,
Major General,
The Adjutant General.

DISTRIBUTION:

B and H (5); Bn 1 (20); C (20).

(For explanation of symbols see FM 21–6.)

TABLE OF CONTENTS

BASIC FIELD MANUAL

PROTECTIVE MEASURES, INDIVIDUALS AND SMALL UNITS

(This manual supersedes Basic Field Manual, volume I, chapter 9, December 16, 1938.)

CHAPTER 1

GENERAL

■ 1. Purpose and Scope.—*a.* This manual discusses the measures to be used by individuals and small units for protection against enemy observation and combat intelligence operations and against enemy weapons, both ground and air.

b. The enemy can be expected to use all possible means to obtain information of the strength, location, dispositions, and movements of our troops. He will also use all available means to prevent us from gaining similar information about his troops. He will seek information by the use of fifth columnists and spies, by observation from the ground and the air, and by scouts and patrols. He will use every available weapon to stop the movements of individuals and units in the combat zone, in the rear areas, and in the zone of the interior.

c. To do your job in the Army effectively, you must know thoroughly the proper methods by which you can conceal yourself from enemy observation and protect yourself from hostile weapons. Only by the correct employment of this knowledge can you efficiently do your job and decrease the power of the enemy to do his.

d. War today puts a higher premium than ever before on individual initiative. It is the action of the individual or small group of individuals that more and more decides the course of modern battles. The intelligent soldier, skilled in the measures which will protect him and his unit against hostile observation and action, will be able to use to the utmost his courage and enterprise in playing his part in the destruction of the enemy.

e. The protective measures which you should know include the use of natural and artificial means of concealment, the use of cover, and other means of protection against air, armored, chemical, and ground attack. You must know how to conceal yourself, your weapon, and your vehicle, horse, or pack animal, how to live and move in terrain which is under hostile observation or fire, how to defend yourself against low-flying aircraft and against armored vehicles, and how to protect yourself against chemical agents. You should know how to perform the duties of a scout in order to reconnoiter or to gain information of the whereabouts, movements, strength, and dispositions of the enemy. Finally you must know how to avoid giving information to the enemy through letters or conversation, or through your answers to questions in the event that you are taken prisoner.

CHAPTER 2

CONCEALMENT

SECTION I

GENERAL

■ 2. GENERAL.—*a*. If an enemy knows the strength of our forces, where they are located, and what they intend to do, he can make plans to use his own forces in such a way that he will defeat our aims or, at the least, make us pay the greatest price in loss of men and material to accomplish our mission. How can he learn these things? He can get a little information from spies and similar sources, perhaps, although such information can never be conclusive. Most of the information he will get, including the confirmation of information from spies, will come from what he sees from the ground and from the air.

b. It is obvious that if an enemy observer sees you, either from the ground or the air, you are giving him information. What is not so obvious is that if he sees, either by eye or from the study of airplane photographs, the marks you have made on the ground, he may learn as much as if he actually saw you make them. To keep the enemy from gaining this information then, you must know how to conceal yourself and how to conceal or disguise the marks you make.

c. Your first duty, always, is to accomplish your mission. If you can remain concealed and leave no marks while accomplishing your mission, whether it be bringing up a truck load of ammunition, preparing a meal, outflanking an enemy machine gun, or anything else, the probability of your success and that of your unit and our Army will be increased.

■ 3. PURPOSE OF CONCEALMENT.—Concealment is a most effective means of protecting you or your unit against enemy observation and action. While you are not necessarily protected from hostile fire when you are concealed, you are protected from aimed fire. You conceal yourself not only to prevent the enemy from stopping you in the accomplishment of your mission but also because you endanger others by exposing yourself. If you neglect to use concealment you may be seen and hit. Your unit not only loses your services but its position is unnecessarily disclosed and others may become casualties. Hence the knowledge and use of concealment measures are needed by every soldier so that he may better play his part as a member of the team.

■ 4. WHAT TO CONCEAL.—To understand the reasons for and to make intelligent use of both natural and artificial means of concealment you should consider those things which disclose the presence of individuals, units, weapons, or equipment to hostile ground or air observers or to the camera. These are movement, reflection of light from bright objects, unnatural colors, regularity of outline, and change in the normal appearance of natural objects or of terrain which the enemy has previously observed.

a. Movement.—Anything in motion instantly attracts the eye; therefore, movement most readily reveals your position. You may be perfectly concealed when motionless but be easily detected when you move. All unnecessary movement must be avoided and when it becomes necessary to move you should do it rapidly, silently, and when your mission permits, from one concealed position to another. (See sec. III.)

b. Reflection of light.—You should expose nothing which glistens. Metal parts of your equipment and weapons are blued to prevent them from reflecting light. They must be kept so. When you are using a watch, compass, or field glasses, shade them from the sun and conceal them when not in use. If you have a vehicle be sure that it is so parked that no light will be reflected from glass or bright parts.

c. Colors.—(1) Colors which do not correspond to those found in nature are quickly observed. White and other bright colors are unnatural under ordinary conditions and exposure of objects of these colors will disclose your presence. Your uniform is of a color which blends naturally with the surroundings under most conditions. In the winter, however, when snow is on the ground, white garments will assist in concealing you. It is rare, in nature, to find any large areas of a solid, unbroken color. Although your tent, truck, or other large object is colored to blend with the colors of nature, you must break up the expanse by some means to prevent the solid area of color from attracting attention.

(2) Under the subject of colors you must also consider lightness and darkness of objects, since the camera records things in terms of black, white, and various shades of gray. These effects are the result of the light reflections of the objects which are photographed. Light rays bounce off surfaces just as a rubber ball does. If the surface is smooth and level it will appear light on the camera film or plate, no matter what the color to the naked eye. For example, a black top road will appear quite white in an airplane photograph. Grass or other vegetation which has been trampled by foot or rolled flat by vehicle wheels shows up lighter in an air photograph than the surrounding vegetation. Light is reflected at many different angles from rough or broken surfaces so that even if they are of a light color to the eye, they appear dark on the film.

d. Regular outline.—You have observed that there is little regularity in the outline of natural objects or of the shadows they cast. Where you find straight lines and uniform spaces on the ground it is almost invariably the work of man and hence discloses his presence. When you are seeking to conceal an object you must break the regular outline of the object itself and of its shadow. Shadows are among the darkest things shown by the camera and the study of shadows is of great assistance in the interpretation of air photographs. Weapon emplacements, parked trucks, trenches, fox holes, and tents must be irregularly spaced even when they are concealed.

e. Changes in terrain.—(1) Every human activity, from taking a short cut across the lawn to building a factory, leaves a visible mark on the face of the earth. The marks which those in the Army make are different from those made by man in his civilian pursuits. Persons in the Army live differently, move differently, and work differently from civilians, so that marks they make are distinctive and easily identified. The marks made by tanks, trucks, artillery, trenches, and all other Army activities differ from each other. If these marks are obvious to the enemy, either through an observer or from the study of air photographs, he will be able to read them easily, discover the strength, dispositions, and intentions, and make his plans accordingly. These marks are not made by some "person or persons unknown" but by each of you, every time you drive a truck or walk across an open field, or dig a hole in the ground, or cut a tree.

(2) When a unit moves into an area that has been under enemy observation, either from the air or the ground, care must be taken to change as little as possible the normal appearance of objects in that area. Four new bushes on the edge of a woods, appearing overnight to an observer or discovered by comparison with a previous photograph of the area, will disclose a battery position to an intelligent enemy, no matter how natural the bushes may appear. If you chop down trees or change their outline by cutting off large branches, the enemy will know that someone is probably in the area so changed. Fresh vehicle tracks leading into a woods will give evidence of the use of the woods, although every vehicle in it may be perfectly concealed. Your aim must be first to make no unnecessary changes in the appearance of the ground, and then either to hide the marks you must make, find a way of making them less conspicuous, or disguise them so they will not disclose valuable information to the enemy.

■ 5. ENEMY OBSERVATION.—There are two hostile sources of information which you must defeat by concealment: the eyes of the enemy ground observer and the eyes and camera of the hostile air observer. Bearing in mind those things which reveal your presence or activity to the enemy, you must consider the terrain from the enemy's viewpoint and make full use of the concealment offered by such features as woods, covered routes, buildings, and other natural objects.

■ 6. CONCEALMENT FROM GROUND OBSERVATION.—You are familiar with how things look at ground level. Concealment from ground observation is primarily a problem of individual concealment or, at most, concealment of relatively small groups of men and materials. You know those things which will reveal your presence to the ground observer. When you learn to make full use of trees, brush, rocks, grass clumps, slopes, banks, and small irregularities in the ground it will be difficult for a ground observer to see you. When you add to these the use of natural and artificial camouflage you will be able to hide from ground observation if your mission permits it.

■ 7. Concealment From Air Observation.—Most of us are not familiar with the air observer's point of view. Hills and valleys, for example, which hide you from a ground observer mean little or nothing to the observer in the air. Tracks, trenches, and ground litter are rarely visible to the enemy on the ground but they all tell a story to the air observer. Everything is spread out as upon a plate for the air observer to look at. He can photograph the ground where he wishes and the photographs can be studied calmly with instruments so exact that much that would escape his naked eye can be discovered. To conceal yourself from him you must use all the measures you use to hide from the ground observer plus all possible care to make no unnecessary marks on the ground, and to conceal or disguise those which must be made. Many of the natural objects of the terrain which will conceal you from ground observation will also conceal you from the air observer. When natural concealment is lacking or is insufficient you must know how to use camouflage.

■ 8. Camouflage.—Camouflage is a word used to refer both to the materials used and the work done to provide protective concealment of military objects from enemy observation.

■ 9. Camouflage Materials.—Either natural or artificial materials or combinations of both may be used for camouflage.

a. Branches, brush, or other vegetation which you obtain from the immediate vicinity of your position are usually the best materials. You must place such material in the same relative position as it is found in nature; trees should be placed upright, branches generally horizontal, and grass suspended vertically. The under side of most leaves is of a different color or texture from the upper side. If you expose the under side of leaves to observation, they will appear unnatural and thus draw attention to the area. If you use branches for camouflage place the bare stumps or cut ends where they will not show. Be sure that the leafy ends extend beyond the edges of the weapon, truck, or other object which you are concealing so that its outline and shadow will be broken. Natural materials are subject to wilting and should be replaced with fresh material as often as becomes necessary. You can determine the need for renewal by comparing the appearance with that of the surrounding growth at approximately 2-hour intervals during daylight. You should also be prepared to renew such material just before daylight. Overnight wilting and the displacement of camouflage material caused by the movement of personnel and equipment in the dark make camouflage errors most apparent at and shortly after daylight. Hide the wilted material where it cannot be observed.

b. Artificial materials which you may have occasion to use are fish nets, chicken wire, burlap or burlap substitute, paints, and lumber. Such materials may be used alone or to supplement natural materials. You will seldom have to use any artificial camouflage but the fish net, except under the supervision of an officer or noncommissioned officer.

■ 10. FUNDAMENTALS.—*a.* You use camouflage to hide or disguise yourself from the enemy to mislead him as to your position, strength, and intention, and confuse him so that he wastes his efforts against you. Whenever you occupy a position you should, if possible, camouflage it before or as soon as you occupy it. Camouflage of works which the enemy has had a chance to observe will attract his attention to the fact that someone is probably still there. However, camouflage of works which the enemy has observed is of value, particularly in areas beyond the scope of his ground observation. Even though such installations have been pinpointed on maps, if they are well camouflaged they will frequently be very difficult precision targets for attack by hostile bombardment and by artillery fire directed by air observation.

b. Camouflage should not only conceal from the view or the camera of the enemy the object which you are hiding but should also disguise the fact that anything has been camouflaged.

c. You must avoid using too much material. Even when you use natural materials, be careful not to use too much since that will make the object and its shadow appear darker than its surroundings, will attract the attention of a hostile observer or show up in a photograph, and thus defeat the purpose of your camouflage.

d. When you have completed your camouflage work get out in front of it, if possible, and look at it from the enemy's point of view. This is the surest way to check the results of your work.

■ 11. PERSONAL CONCEALMENT.—*a. General.*—(1) Your ability to conceal yourself involves your recognition of the presence and value of all forms of natural concealment and camouflage, and your knowledge and use of the proper methods of taking advantage of them. Your mission will rarely permit you to hide yourself completely from the enemy since you must observe, fire, or move. You must seek always to become so nearly a part of your surroundings that your location will be unknown to the enemy, while at the same time you will be able to see and act against him.

(2) You may improve your concealment by the use of natural materials common to the area about you. Twigs, weeds, or leaves fastened to your helmet by rubber bands, twine, or a piece of vine, and small branches in your belt will help you blend in with surrounding undergrowth. Such material, however, will reveal rather than hide your position if it differs from the surrounding terrain (fig. 1). You must always remember that materials will appear natural only if the same materials are to be found growing close about you (fig. 2).

(3) Learn to consider the effect of the background in concealing or exposing your location. If you offer a distinct contrast to the background you will be quickly observed. Remember also that the ground beneath you is the background from the air observer's viewpoint. Never silhouette yourself against an unfavorable background or the skyline if you can avoid it (figs, 3 and 4).

(4) Proper use of shade and shadows will help conceal you (fig. 5). Heavy shadows in the middle of the day offer excellent concealment. In the early morning and late in the afternoon shadows are light and usually afford little if any concealment (fig. 6). The long shadow which your body casts in the evening may disclose your presence. If you are forced to move in the open when shadows are long, keep your body as low as possible (fig. 7). If you use shadows to conceal a vehicle or weapon, remember to change its position during the day as the shadows change (fig. 8).

① WRONG—Leafy camouflage against an improper background *discloses* your position.

② RIGHT—Camouflage which blends into the background *protects* you.

FIGURE 1.

① WRONG—Camouflage material alone is useless if it attracts attention.

② RIGHT—Select an inconspicuous location. Blend your camouflage with the natural growth.

FIGURE 2.

① WRONG—Contrasting background will disclose your location even when you are in heavy shadows.

② RIGHT—Favorable background is always important, especially when you are moving.

FIGURE 3.

① WRONG—Purpose of special clothing is lost with wrong background.

② RIGHT—Your clothing should blend with your background.

FIGURE 4.

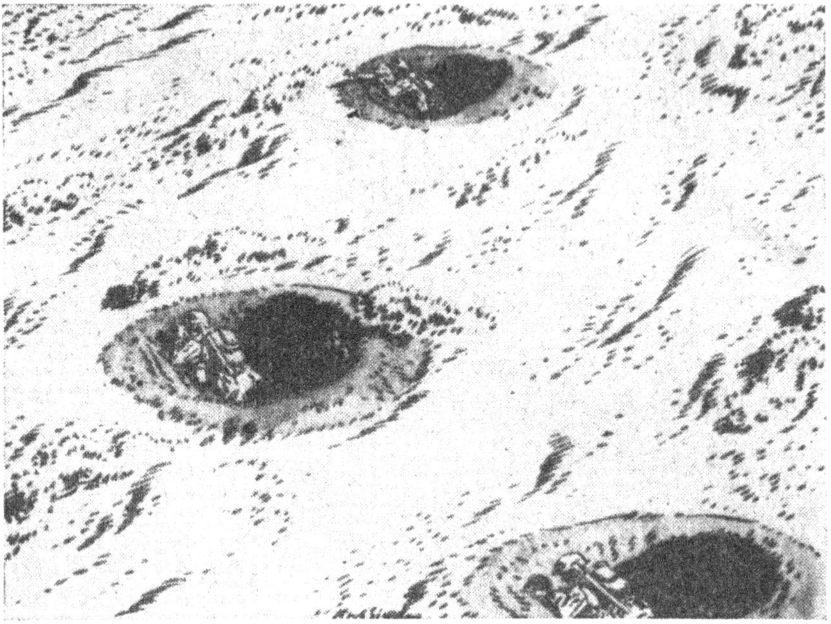

① WRONG—You can be seen from above if you are on the sunny side of a shell hole.

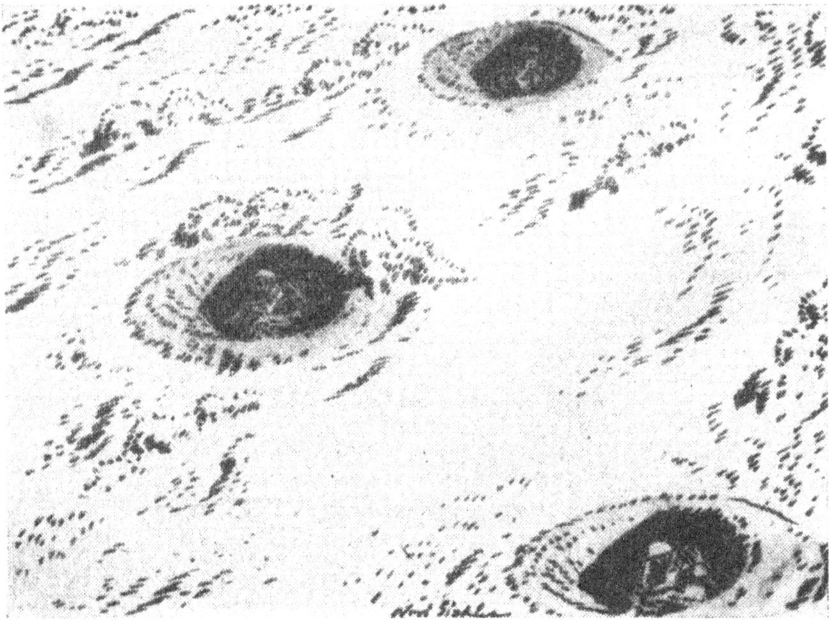

② RIGHT—Shadows aid concealment, particularly from air observation.

FIGURE 5.

① WRONG—Late evening shadows are light and offer little protection.

② RIGHT—Heavy midday shadows afford good concealment.

FIGURE 6.

(5) When you dig a hasty trench or fox hole for cover you should conceal all trace of your work. Remove or conceal the freshly dug earth. A light mat made of brush, branches. or chicken wire net thrown over your fox hole or slit trench and covered with small twigs, leaves, or grass will conceal your location (fig. 9).

(6) When you are going to fire from a concealed position behind a tree, rock, house, or similar object you should habitually fire around the right side of it as you expose less of your body that way. If you are only going to observe, observe from the shady side to avoid casting a shadow which might reveal your position. Don't look over the top of concealment unless it has a broken or bushy top line and the background is broken. Observe through the bush if possible. (See fig. 46.)

(7) Concealment in snow-covered terrain presents a special problem. Ordinary uniform clothing will make a sharp silhouette; shadows and tracks will show up very clearly. You will be provided with clothes or a cape of an off-white color. Shadows and tracks are much more apparent under these conditions and you will have to take great care to prevent them from disclosing your position. In order to provide better observation and concealment from a fox hole or weapon position, use cloth of a coarse weave, such as mosquito netting, rather than smooth cotton sheets or similar materials which reflect light and thus attract attention. If you must occupy a position for some time and cannot avoid making tracks, don't let the tracks end at your position but make them run by it to a dummy position or until they join some other tracks.

(8) The use of individual concealment is more fully discussed in chapter 3. You should be familiar with the important duty of scouting and with the protective measures which a scout must use to accomplish his mission. These measures are applicable in varying degree to every soldier when he is under enemy observation or fire.

b. Rules for individual concealment.—For individual concealment remember to—

(1) Make full use of all natural cover and concealment. The enemy can't aim at you if he can't see you.

(2) Choose your position carefully. Concealment from both ground and air is necessary. Consider the position from the enemy's point of view.

① WRONG—Long shadow cast when sun is low helps enemy locate you.

② RIGHT—Avoid casting shadow by keeping low.

FIGURE 7.

① RIGHT—Shadows help conceal your truck.

② WRONG—Don't forget to move as shadow shifts.

FIGURE 8.

① When you dig a fox hole, conceal the dirt.

② Light mat of shrubs and grass will conceal your fox hole.

FIGURE 9.

(3) Avoid unnecessary movement. When you do move, pick out a concealed place to which to move, think how you can move there without attracting enemy attention, and then move quickly.

(4) Use ditches, hedges, edges of woods, and folds of the ground. These accidents of the ground will prevent accidents to you.

(5) Avoid outstanding landmarks such as lone trees or rocks, fence corners, etc. Such points are easily picked up as targets by enemy observers.

(6) Keep in shadows; they are good substitutes for other concealment.

(7) Smear your face with dust, mud, or grease paint. Your face may be your misfortune if it is bright and shining.

(8) Remove or conceal any equipment which reflects sunlight.

(9) Use extra care when you are tired. Fatigue leads to carelessness.

■ 12. BIVOUACS.—*a*. When you put up any sort of tent or shelter, it should be placed so that it will be in dense shadows both in daylight and when airplane flares are used at night. Such shelters must never be placed in lines or other regular pattern. No tents should be left up during daylight unless they are thoroughly concealed. You should always camouflage the canvas surfaces and also the openings where dark shadows would otherwise be evident. Branches of surrounding foliage, loose leaves, grass, or other local debris will be excellent materials for this purpose (fig. 10).

b. You must be particularly careful in bivouac not to leave any articles such as underwear, papers, blankets, or equipment exposed on the ground, even at night. Bury or otherwise conceal your empty emergency ration and other tins to prevent them from reflecting light which would indicate activity in the area.

■ 13. VEHICLE CONCEALMENT.—*a*. Vehicles present a difficult concealment problem because of their size and regular outline. The great increase in the use of motor vehicles and the value to the enemy of their location, both for information and as targets, make it most important that every sol-

① Use natural materials to camouflage your tent.

② Make it appear to be a part of its surroundings.

FIGURE 10.

dier know how to hide them. A vehicle is conspicuous to a hostile observer because of its rectangular shape, its shadow, and the tracks it makes. Nothing can hide a moving vehicle on a road if the road can be seen by a hostile observer. You can conceal a parked vehicle if you place it in a concealed location or use natural or artificial camouflage to prevent its shape, shadow, and tracks from revealing its location.

b. If you are in charge of a vehicle which must be concealed, pick out a spot which offers some natural concealment from observation, if one is available, and where you can use camouflage to blend in with the surrounding terrain. A place in the shadows cast by trees or shrubbery is excellent (fig. 11).

c. If an open field is all that is available, don't consider concealment impossible. If you park in the center of the field, camouflage won't hide the vehicle, for not only will the unusual object in the otherwise smooth field attract attention, but the tracks will point to your hiding place. Pick a place along one of the edges of the field where the shadow of your vehicle will fall in the shadow of low brush, a hedge, or a ditch. If buildings are near, you may hide your vehicle in one, or you can make your vehicle seem a part of the existing pattern and appear as a lean-to addition to a building (fig. 12).

d. Determine the best concealed approach to the spot you have selected. Remember that good concealment may be spoiled by carelessly made tracks. Move the vehicle off the road and to the selected point so that, if possible, the tracks are concealed or may be erased, and park it headed in the direction of probable departure. If visible tracks must be made, make them intelligently. Do not cut across a field to the parking place (fig. 23). Follow a fence or hedge or furrows. Vehicle tracks are very apparent in soft or grass-covered ground and remain visible for a long time.

e. When you camouflage the vehicle you must break up its outline and the shadow it casts (fig. 13). Natural materials are best, if available, and you must use care to place the material as it grows in nature. Don't use too much material or the appearance will be unnatural. Although the vehicle is painted an inconspicuous color, it has a sharp, unnatural outline which should be broken.

① WRONG—Do not forget that a position in morning shade will be in afternoon sun.

② RIGHT—Intelligent use of shade affords concealment.

FIGURE 11.

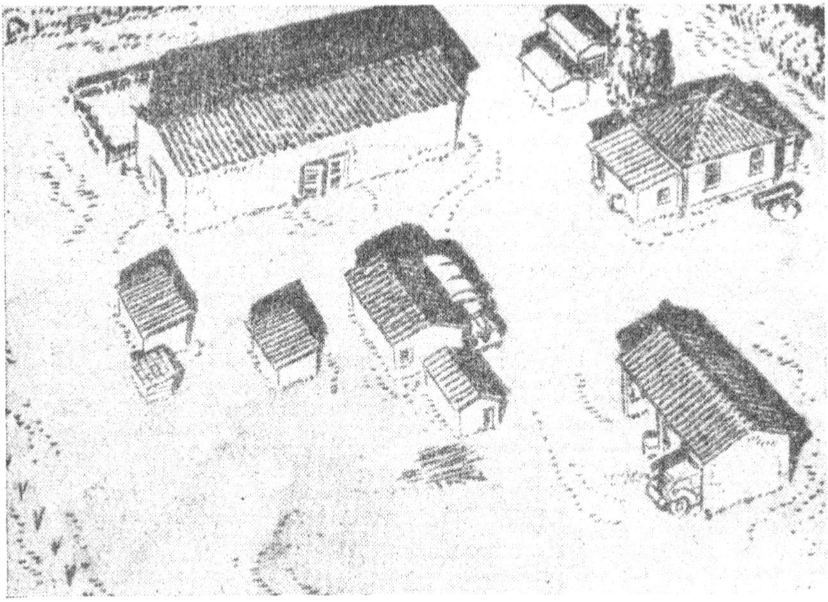

① WRONG—Military vehicles near buildings attract attention.

② RIGHT—Make your vehicle appear to be part of the surroundings.

FIGURE 12.

① WRONG—Camouflage is not effective when it leaves outline of object distinct.

② RIGHT—Eliminate outline of object *and its shadow.*

FIGURE 13.

f. Except in barren areas, such as a desert, you will ordinarily be able to secure natural camouflage material in adequate quantity. However, if you can't find sufficient natural material, a camouflage net should be used. A net should have a mesh of about 2½ inches and be of adequate size for the vehicle. Such a net is not a magic cloak which will make all objects beneath it invisible. If it is placed over the top of the vehicle it will not make the rectangular shape and the shadow disappear. The net is used as a curtain and a base to hold garnishing, both natural and artificial. The top of the net should be tied to some support above the vehicle, a tree or pole, for example. Then the edges of the net should be pulled away from the vehicle and pegged to the ground (fig. 14).

g. The net must be garnished either with natural or artificial material. Small branches and twigs can be drawn through the openings in the net so as to be held in an upright position as they grow in nature. Garnishing is more often done with artificial materials and usually consists of strips of cloth interwoven in irregular patterns, the ends left hanging to cast broken shadows. These strips of cloth are dyed to match the local foliage. This type of garnishing is usually done before the net is issued for use. Sacks or strips of burlap or old canvas may sometimes be used. In the vicinity of buildings, old lumber may be used to break the outline and shadow a vehicle.

h. Your windshield and headlight lenses will reflect light to a lesser degree if they are faced away from the sun. To cover the windshield and side windows, you may throw a blanket over the cab and hold it in place by catching the edges between the doors and cab frame. Use grease or mud to prevent the reflection of light from other bright surfaces—headlights, for example. In cold weather a stationary vehicle with its engine running may be detected by the white cloud of its exhaust.

i. If you are in charge of several vehicles don't permit them to park in a group, line, or other regular pattern. If practicable, each vehicle should have its own concealment. Make full use of trees and their shadows (fig. 15). Have the drivers remain near the parked vehicles, concealed under bushes or trees. They should be cautioned about moving about and never allowed to gather in groups.

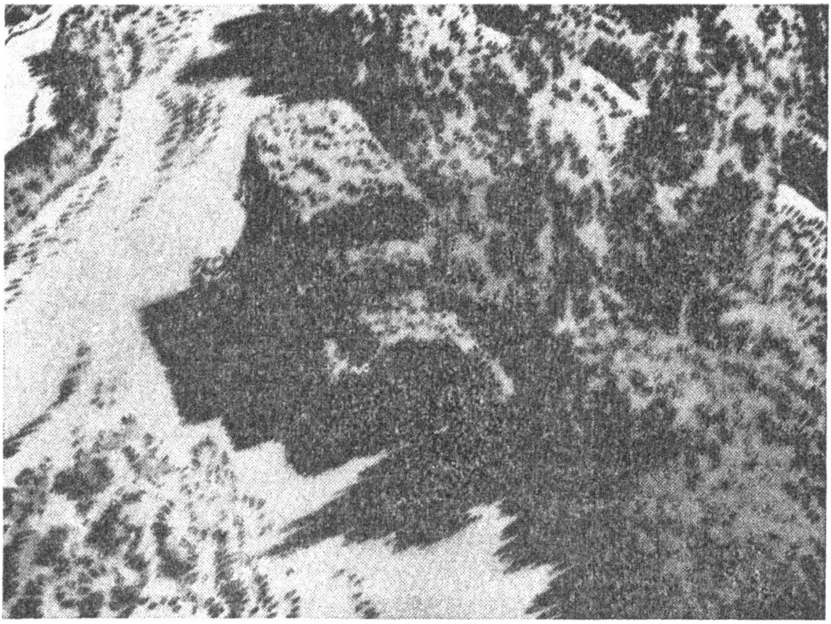

① WRONG—This net hides truck but leaves shadow.

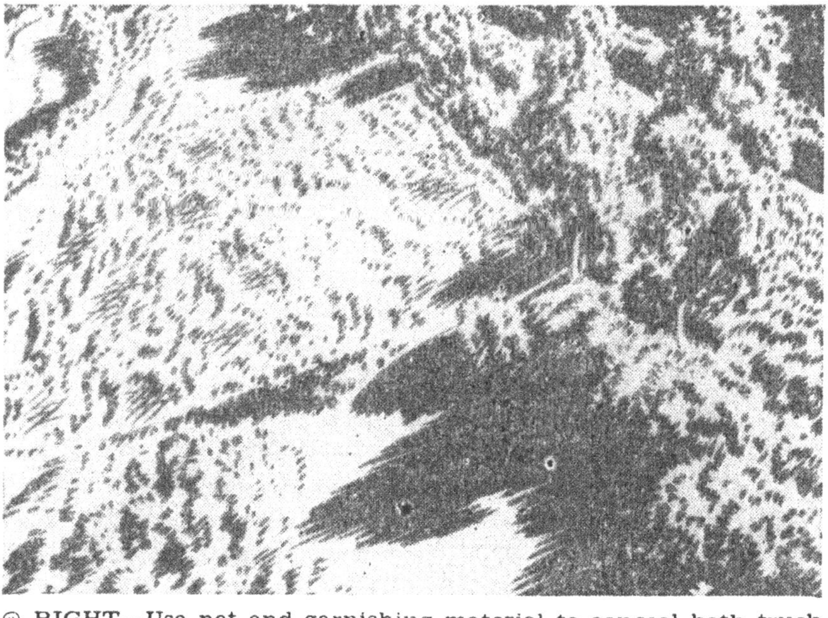

② RIGHT—Use net and garnishing material to conceal both truck and shadow.

FIGURE 14.

① WRONG—Partial concealment affords protection to none.

② RIGHT—Individual concealment of each vehicle increases protection for all.

FIGURE 15.

■ 14. WEAPON CONCEALMENT.—*a.* If you are charged with the concealment of a weapon, a machine gun or antitank gun for example, you have two problems to solve. You must have a good field of fire and the position must be difficult for the enemy to locate. You must avoid an obvious or conspicuous location (fig. 16). The best camouflage will be of little value if the weapon position is in a location that is easily recognized and can be clearly indicated on the ground or a map to an artillery observer or to airplanes.

b. The camouflage material must appear natural in the surroundings (fig. 17).

c. When a gun is fired, the blast of hot gases burns or otherwise marks the ground for some distance in front of the muzzle. The blast mark will often disclose a gun position after a few rounds have been fired. You must provide material for covering the blast mark if the gun is to remain in the position after firing (fig. 18). Such camouflage should be removed, if time permits, before the gun is fired again, or inspected and renewed when necessary after the gun has fired. Wetting down the ground in front of the gun, if it can be done without disclosing the position, will lessen but not eliminate blast marks.

d. The outline of the gun and its armor must be broken and made to blend in with the surrounding ground. In barren ground this may be done by attaching pieces of cardboard, cloth, or wood to the shield (fig. 19).

e. If a sheet or net is needed to conceal the gun it must be made to appear natural, otherwise the artificial appearance of the net itself will disclose the position (fig. 20).

f. Brass cartridge cases, powder bags, ammunition containers, and spare parts packages should be covered or otherwise concealed to avoid attracting attention to the position.

■ 15. CAMOUFLAGE DISCIPLINE.—*a.* All commanders to include squad and half-squad leaders are required to take appropriate action for the camouflage and concealment of their commands. Measures taken by commanders can be of little value unless every individual in the unit gives complete and intelligent cooperation and assistance, since careless action by a single individual may disclose an otherwise carefully concealed unit. The best camouflage material and work will be of no value in gaining concealment without camouflage discipline.

① WRONG—A conspicuous position, though well camouflaged, may arouse suspicion.

② RIGHT—Avoid the most conspicuous location.

FIGURE 16.

① WRONG—Unnatural camouflage betrays rather than protects position.

② RIGHT—Use material that blends readily with surroundings.

FIGURE 17.

① WRONG—Blast marks will disclose an otherwise carefully camouflaged gun.

② RIGHT—Conceal the blast marks whenever opportunity offers.

FIGURE 18.

① WRONG—This camouflage doesn't break the straight lines.

② RIGHT—Camouflage which breaks the straight lines affords concealment even in barren ground.

FIGURE 19.

① WRONG—Shadow cast by this net discloses gun location.

② RIGHT—Arrange net to eliminate shadow and blend with surroundings.

FIGURE 20.

b. Camouflage discipline has two objectives:

(1) Prevention of any change in the appearance of the visible terrain by the personnel occupying it.

(2) Maintenance of camouflage material.

■ 16. CHANGES IN TERRAIN.—*a.* Outside of movement, the first evidence an observer has of the presence of his enemy is a change in the normal appearance of the terrain. The longer the time the enemy has had to observe the terrain the more quickly will he note changes. The most noticeable changes are new paths or tracks, trees cut down or sod displaced, foreign objects, and newly dug earth. You must always be careful not to make such changes without proper camouflage measures and to caution others against doing so. Whenever you can, use existing roads, trails, and paths (fig. 21).

b. If you must make a new path make it under cover of trees or bushes or along a ditch, hedge, furrows, or fence line (fig. 22). If this is impossible obliterate the tracks by sweeping over them with boughs of trees or similar materials. Never end the new path at a camouflaged installation but continue it until it joins an existing road or trail, then backtrack (fig. 23). If time permits, all paths both new and old should be wired or taped at the sides to prevent widening which would indicate heavy use. Don't permit vehicles to turn around at a camouflaged position. Have them move on to a normal turning point.

c. You should never cut or change the appearance of trees which the enemy has had a chance to observe. When sod is cut save it and use it to cover the newly dug earth. Don't throw things around on the ground but conceal or bury them.

d. When you are performing any duty near a concealed location—for example, a messenger reporting to a message center or command post, delivering ammunition to a gun position or rations to a kitchen, bringing an ambulance to an aid station or a maintenance truck to a truck park, use the greatest care not to disclose the location.

e. Whether you are on foot, mounted, or driving a vehicle, follow a used road or path, ask for information, and obey the instructions of sentries posted to enforce camouflage discipline.

① WRONG—Numerous tracks point out the concealed location.

② RIGHT—When tracks must be made, make them as inconspicuous as possible.

FIGURE 21.

① WRONG—Excellent concealment is ruined by numerous tracks.

② RIGHT—Restrict tracks and paths to the cover of hedges, trees, and bushes.

FIGURE 22.

① WRONG—Tracks ending abruptly in a locality will indicate something concealed there.

② RIGHT—When new tracks must be made, continue them past the concealed installation.

FIGURE 23.

■ 17. MAINTENANCE OF CAMOUFLAGE.—Camouflage material is easily displaced or damaged. You should take a good look at your camouflaged position to impress its appearance on your memory. Then inspect it frequently and correct any changes made by displacement or wilting. If you are on duty as a camouflage sentry get a clear mental picture of the camouflage when you are posted and be sure it looks the same at the end of your tour.

■ 18. DON'TS IN CAMOUFLAGE.—*a.* Don't be careless and thus expose your comrades to enemy attacks. They are depending on you just as you must depend on them.

b. Don't look up at airplanes. The enemy is looking for you too, and you're easier to hit than he is.

c. Don't walk in the open. Every time you put your foot down you leave a mark of 48 square inches to attract enemy attention.

d. Don't gather in groups. Dispersion of men insures longer life for all.

e. Don't dry clothing in the open. It acts as a signal to point out your unit to enemy airplanes.

f. Don't throw newspapers, boxes, ration tins, or cans in the open. They indicate activity in the area.

g. Don't use flashlights or strike matches in a combat area at night. They serve the enemy as a beacon does an air pilot.

h. Don't park your truck with the windshield uncovered or the rear flap open. If you attract any visitors they're sure to be of the unwelcome variety.

i. Don't forget that the enemy is just as intelligent as you are. Anything that would disclose his presence to you will disclose your presence to him.

SECTION II

COVER

■ 19. GENERAL.—*a.* Cover is protection against the fire of hostile weapons. The type of cover which you seek will depend upon the kind of enemy fire against which you are seeking protection. A reverse slope that would give you protection from rifle or machine-gun fire will probably not protect you from the high-angle fire of mortars or howitzers. As in concealment, when you are looking for cover study the terrain from the enemy's point of view. Many natural objects will give you cover and, if time and your mission permit, you can use artificial means to obtain or improve cover.

b. You are most exposed to enemy fire when you are standing, much less when you are prone, and best protected when you are below the surface of the ground. For example, under artillery shell fire of ten men who would be hit by shell fragments while standing, only six would be hit if all were prone, and but one if they were in shallow trenches or fox holes. In a machine-gun attack by low-flying airplanes your chances of not being hit are four times as good if you are flat on the ground, even on a flat road, as they are if you are standing. Your chances of escaping injury from bomb fragments are six times better prone than if you are standing. If you are in even a small depression during an air attack there is very little chance of being hit by either machine-gun fire or bomb fragments.

■ 20. Natural Cover.—Many natural objects, trees, rocks, ditches, embankments, and folds in the ground, as well as shell holes, afford protection from hostile fire. Such cover is readily apparent to every eye. You must learn to study the terrain in order to appreciate the cover afforded by the slightest depressions and lumps in ground that appears flat to the untrained eye. By making full use of all natural cover and using the method of movement best adapted to the situation (sec. III) you will have considerable protection while moving under hostile fire. When you are to remain in a position you should use all available time to improve your cover by artificial means.

■ 21. Hasty Entrenchments.—In order to improve natural protection against hostile fire by hasty construction you must have a knowledge of the tools available, their use, and the types of hasty entrenchments which will afford cover. Permanent or semipermanent cover requires a long time and many men to construct, and will normally be done under the supervision of an officer. When your mission permits, you should provide or improve your cover by digging. You must know the various types of cover which you can provide and learn how best to construct them by digging them. These types have been developed by survivors of hostile attacks and tested under fire. It is hard work, and it requires practice to dig them quickly and properly. Learn how *before* hostile action forces you to.

■ 22. TOOLS.—*a*. All dismounted combat troops carry with them small individual tools for making hasty entrenchments. The following such small tools (fig. 24 ①) are carried by individuals in the 12-man infantry squad: 1 wire cutter, 1 ax, entrenching, 3 pick mattocks, and 7 shovels, entrenching.

b. In addition to these small tools, entrenching sets, carried by the engineers, are issued as needed by large groups for digging entrenchments. These sets include, among other tools, the standard size ax, pick mattock, and shovel (fig. 24 ②). Each truck and similar vehicle carries these three tools. You should know how to use them effectively to construct cover.

■ 23. SKIRMISHER'S TRENCH.—When you are under small-arms fire you can secure protection by digging a long, shallow trench while lying down flat on the ground with your head toward the enemy. Use your own individual tools, or your bayonet, mess kit cover, sticks, or any other available object. Lie on your left side and use the tools available to dig a small trench in which your head and body will fit sideways (fig. 25). Roll over into this and, lying on your right side, reverse the position of your hands to widen the trench. When you have cover for your head, shoulders, and hips extend the trench to the rear to provide cover for your legs. Throw the dirt to the front to form a parapet to increase your protection. In average soil you can get fair protection in about 10 minutes and finish the trench in less than an hour. The finished trench (fig. 26) will give you protection against flat trajectory small-arms fire, but only partial protection against high-explosive shell or bomb fragments. You should enlarge the forward portion into a fox hole as soon as enemy fire permits.

■ 24. FOX HOLES.—This is the more usual form of hasty entrenchment. You may start to dig it from a prone or crouching position. You will be able to dig it more easily, however, from a standing position when not under fire. It is developed

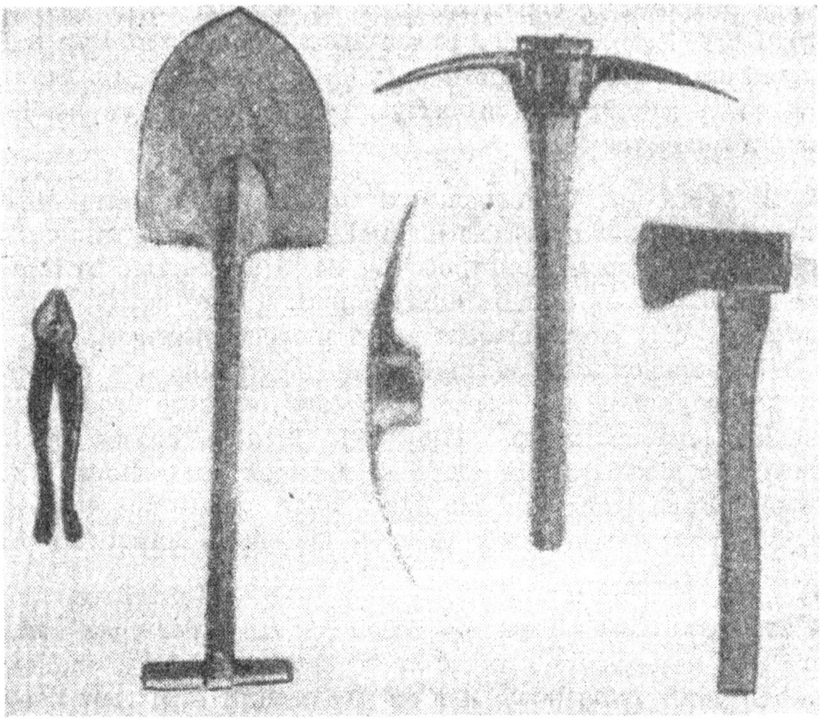

① Individual tools carried by infantry soldiers.

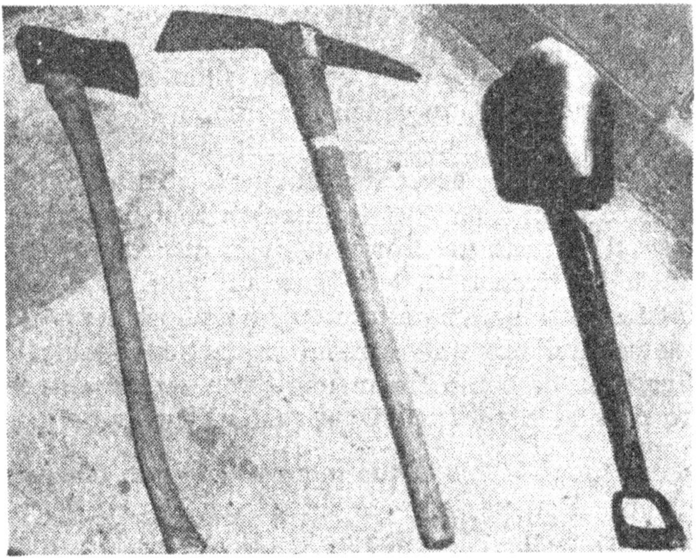

② Standard size entrenching and vehicle tools.

FIGURE 24.

successively from the squatting type to the kneeling and finally to the standing type (fig. 27). In average soil you can dig the standing type, using standard size tools, in 1 hour or less; with individual infantry tools, in about an hour and a half. This type of entrenchment will give you satisfastory

FIGURE 25.—Use of individual pick and shovel, prone position.

protection against small-arms fire and the fragments of artillery shell and bombs. When water collects dig a small hole (sump), larger than a canteen cup, at the lowest part of the bottom to facilitate bailing. In firm soil, with the lower portion enlarged, you can curl up in the bottom and secure better protection from shell fire, and protection from tanks passing overhead.

■ 25. SHELL HOLES.—With a small amount of labor you can improve a shell hole to obtain quick protection and good concealment (fig. 28). Such improvement consists of a recess dug 2 or 3 feet into the forward slope to get a good firing position and lateral protection from shell fragments and enfilade fire. If you are taking cover in a ditch or behind a bank, a similar recess dug in the edge will furnish you considerably more cover than the unimproved bank or ditch.

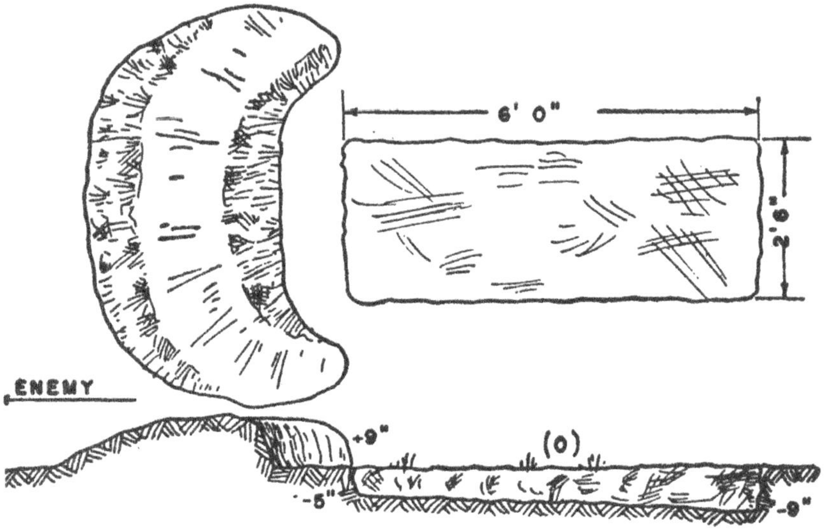

FIGURE 26.—Skirmisher's trench.

■ 26. SLIT TRENCHES.—*a*. This type of trench has become very important in the present war. It gives excellent protection against all types of fire, air attack, and, in firm soil or when revetted in soft soil, provides protection against tanks passing overhead. It is an excellent type of cover for the immediate protection of gun and vehicle crews and for antitank lookouts.

b. A slit trench is less visible to ground observation if it is dug parallel to the front and the spoil (dug out earth) scattered and concealed rather than used as a parapet. The cut sod should be saved and used for camouflage. Such a trench can be concealed by methods similar to those used in

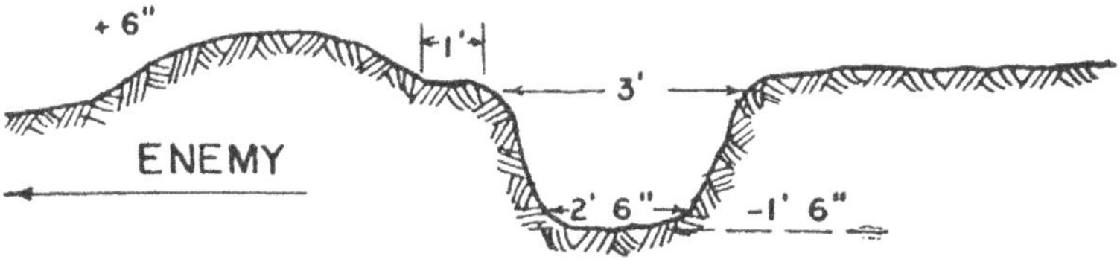

① Squatting type (first stage).

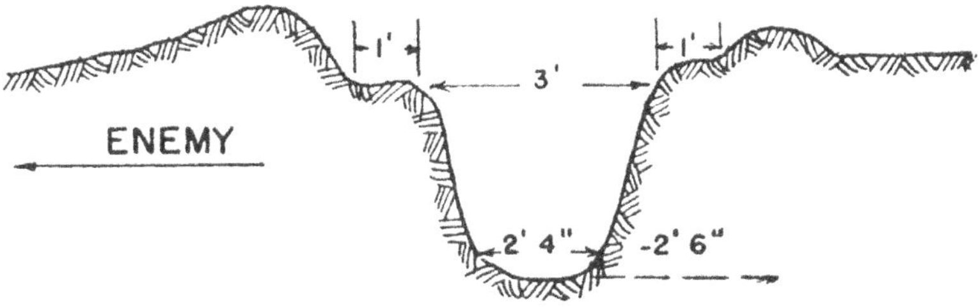

② Kneeling type (second stage).

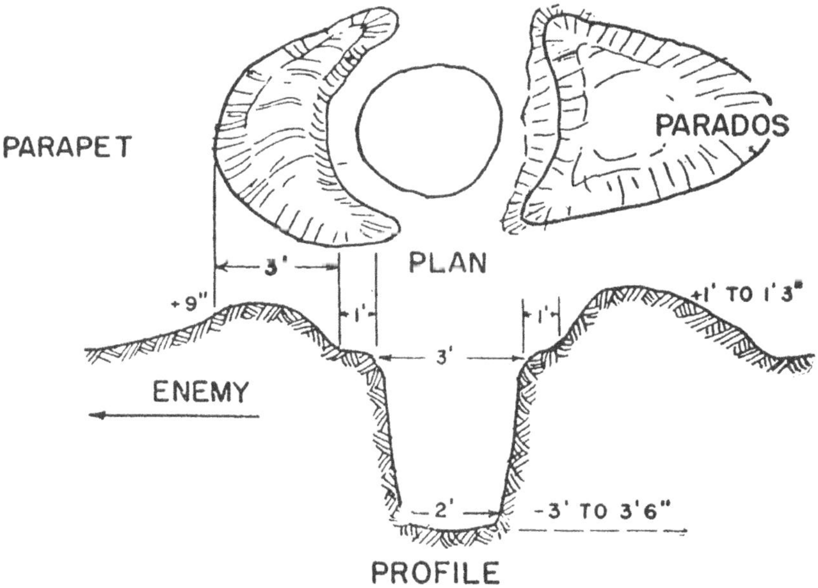

③ Standing type (third stage).

FIGURE 27.—Development of fox holes.

camouflage of a fox hole (fig. 9). A slit trench should be as narrow as possible and still admit you, and deep enough to permit you to get below the surface of the ground. A standing slit trench may be caved in by concentrated artillery fire. For this reason one dug in soft ground should

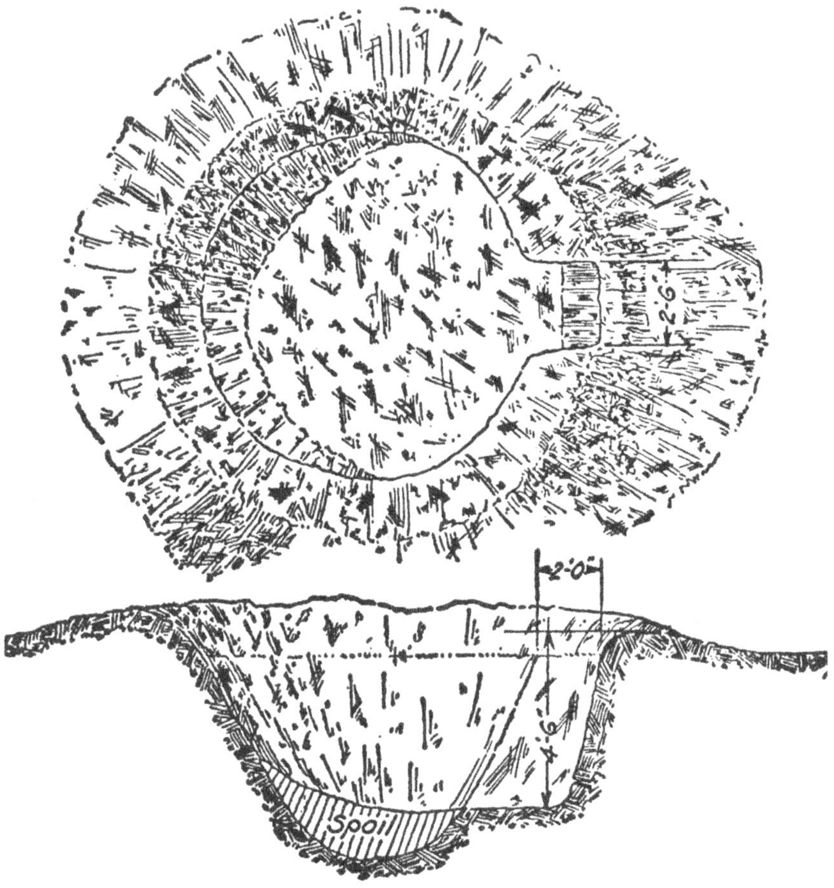

FIGURE 28.—Improvement of shell hole.

be well braced and revetted. A single such trench should not be required to hold more than two individuals. When more are to be protected, dig more trenches. Slit trenches in the shape of a chevron or cross, about 6 feet on a side, will assure protection against enfilade fire from tanks.

■ 27. SHALLOW CONNECTING TRENCHES.—After individual cover has been provided a shallow trench may be dug con-

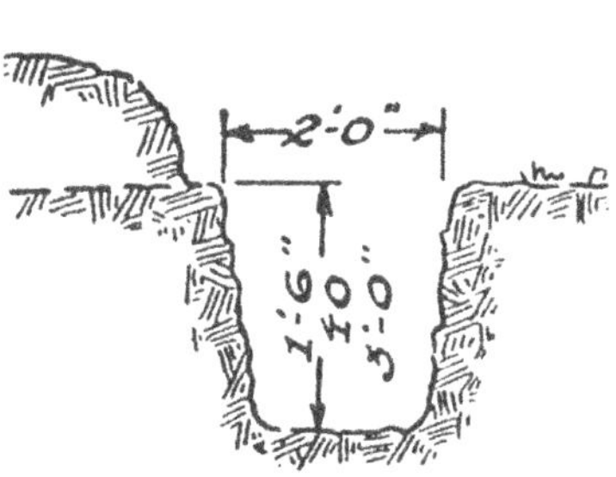

① Creeping.

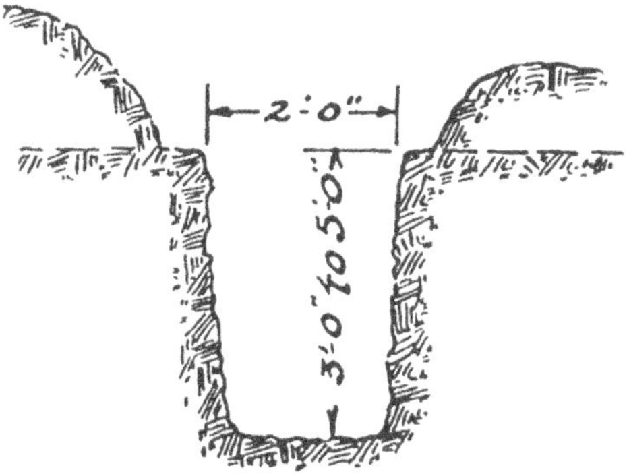

② Standing.

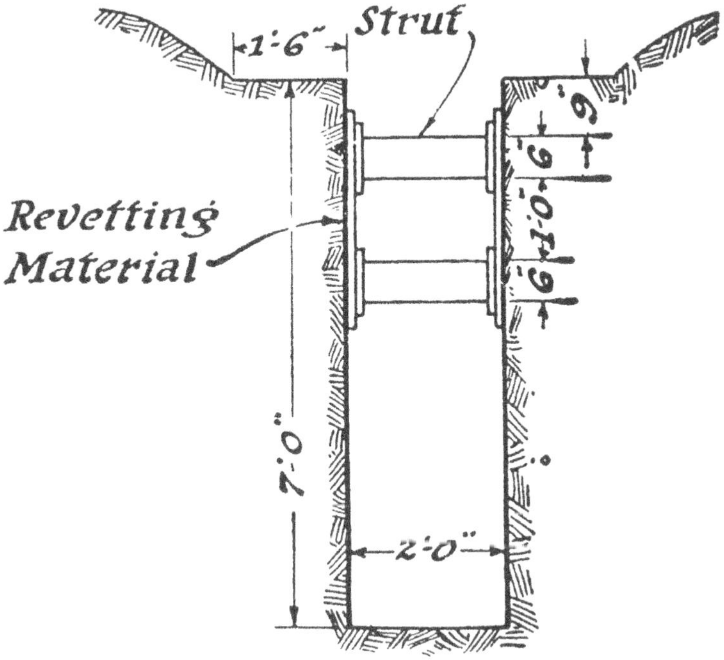

③ Deepened, to protect against shell fire.

FIGURE 29.—Types of slit trench.

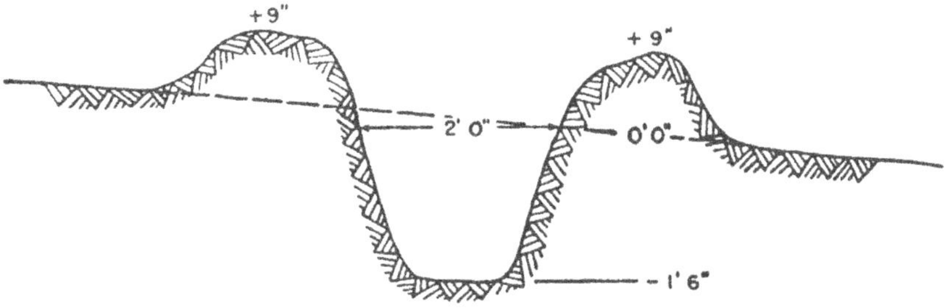

FIGURE 30.—Shallow connecting trench.

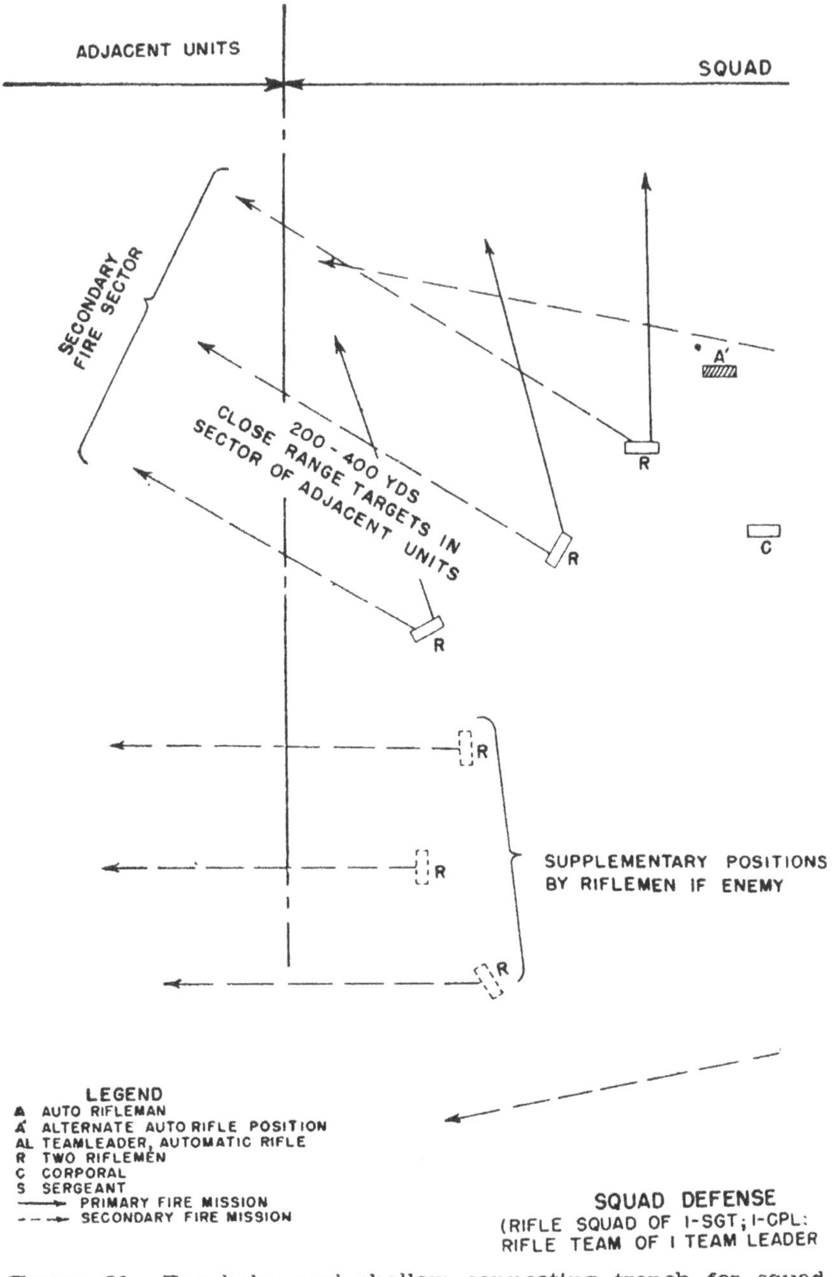

FIGURE 31.—Fox holes and shallow connecting trench for squad. position.)

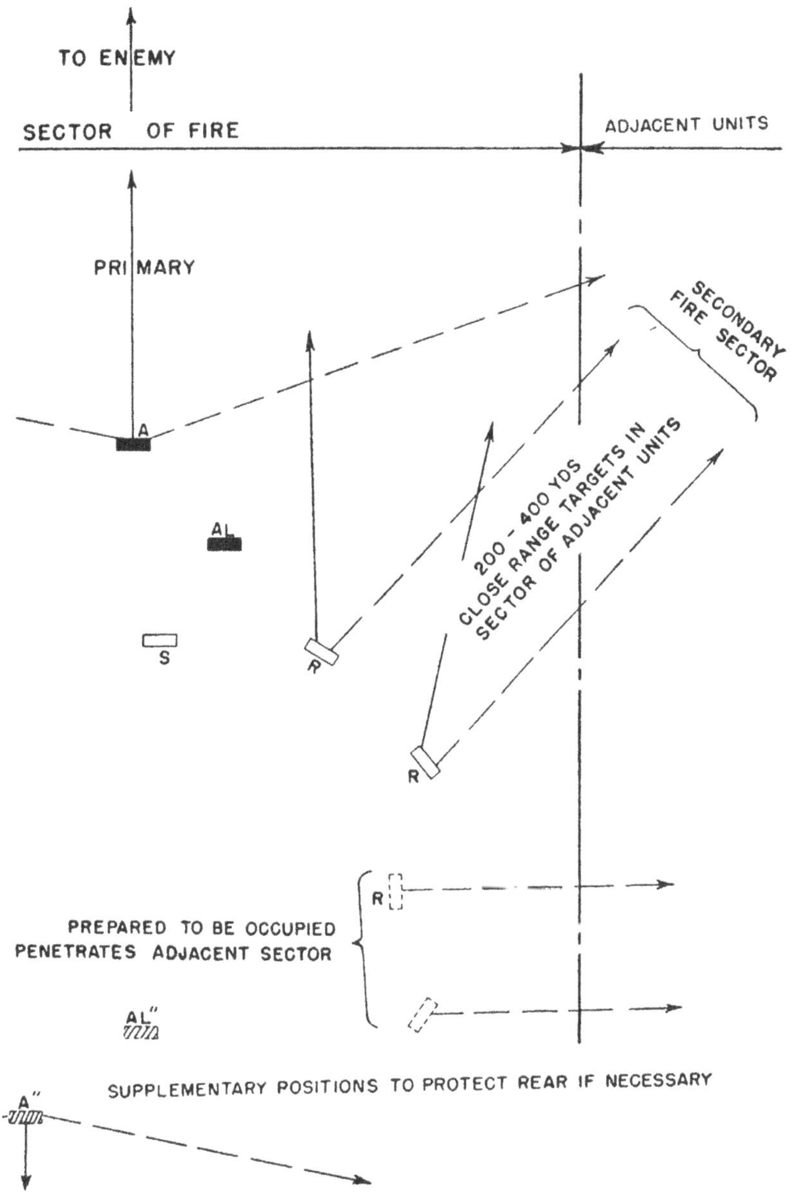

AREA— USING TWO-MAN FIRE SLITS.
10-PVTS, WITH 3-MAN AUTOMATIC
AND 2 AUTOMATIC RIFLEMEN.)

(Shows primary and secondary fire sectors and supplementary

necting individuals (fig. 30). You may use this type of trench for lateral communication between fox holes or slit trenches. It will provide cover when you are crawling on your hands and knees.

■ 28. SQUAD POSITIONS.—*a*. If you are preparing your squad for defense, you should deploy your men and assign to each man a primary position and a primary and secondary sector of fire to meet the most probable direction of attack. Since all-around defense is a necessity for every unit, you must also designate any supplementary positions and secondary fire sectors needed to protect the flanks and rear of your squad position. Primary fire sectors cover the squad sector of fire toward the enemy. The secondary fire sectors cover areas at close ranges, 200 to 400 yards, to the flanks, in the sectors of adjacent squads. Supplementary positions are prepared for protection against attack from flank or rear in case of hostile penetration of adjacent sectors (fig. 31).

b. Fox holes for each primary and supplementary position are started as soon as possible after you deploy your squad, and more fully developed as time and the situation permit. Individual fox holes should be about 5 yards apart or they may be placed in pairs. If the position is to be held for some time, have the fox holes connected where necessary by shallow connecting trenches (par. 27). If your men are to occupy the fox holes overnight, have them extend the fox holes on each side or deepen the connecting trenches so they can lie prone while sleeping.

c. If an automatic weapon, automatic rifle, machine gun, or submachine gun is available you should site it in an advanced position near the center of your group of fox holes so that its fire can cover the entire fire sector of your squad and the fronts of adjacent squads. Select an alternate position nearby to which it can move, if necessary, and deliver the same fire. Select a secondary position to permit its fire to cover the rear.

SECTION III

MOVEMENT UNDER ENEMY OBSERVATION OR FIRE

■ 29. GENERAL.—Movement is more apt to attract the attention of enemy observers and to draw enemy fire than any other thing you do. A sudden jerk of your arm, the quick movement of a weapon, or the puff of your breath on a cold morning will catch the eye and focus observation on your position. Yet movement is necessary in combat in order that you may close with the enemy and destroy him. You must develop your self-control to such degree that you make no unnecessary movement, and train yourself to move rapidly and in such a manner as to attract the least attention and to present little or no target for hostile fire.

■ 30. UNNECESSARY MOVEMENT.—In the presence of the enemy you must remain motionless except for such move-

FIGURE 32.—Prone positions.

ment as is necessary in firing, observing, transmitting orders or signals, or attacking. When you stop in the open or behind any slight concealment or cover and your unit is not actually firing at the enemy, you muse *lie motionless in the prone position,* with your body stretched out as close to the ground as possible. Your body should be flat, left cheek on the ground, legs extended and spread out. Your heels should be turned in and touch the ground. Extend your arms to the front, flat on the ground. If you are armed, grasp your rifle in your right hand at the balance (fig. 32.)

■ 31. INDIVIDUAL MOVEMENT.—You may advance by running, creeping, or crawling. You must learn to execute the movements with precision, vigor, and minimum exposure.

■ 32. RUNNING.—You will advance by running when the squad advances by squad rushes or when advance is by indi-

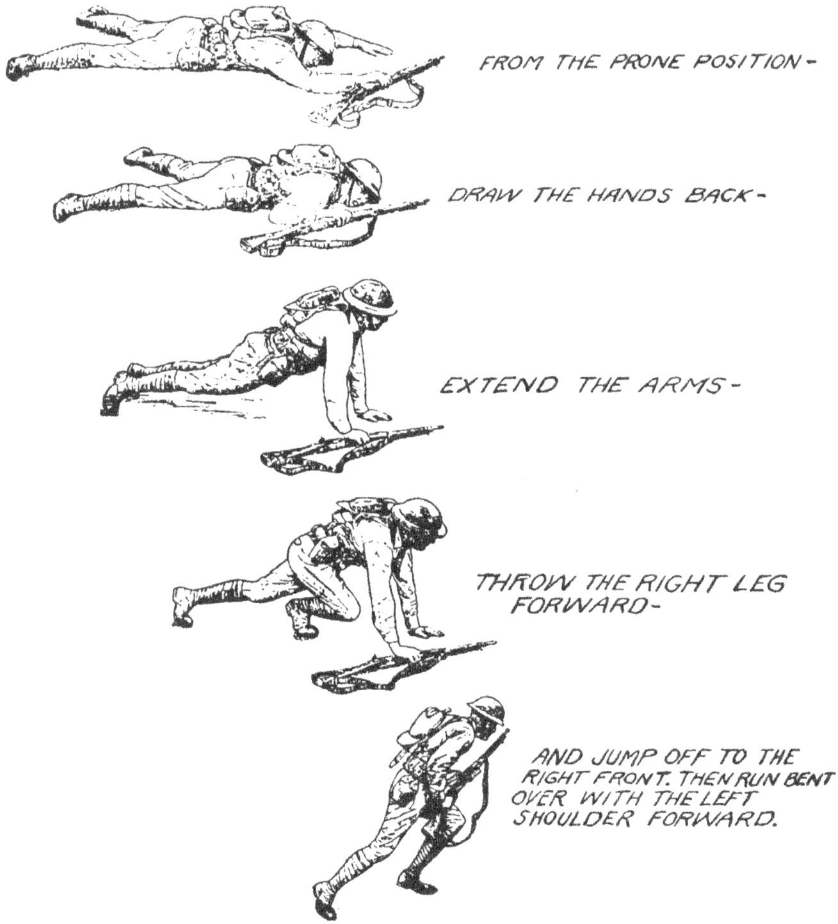

FIGURE 33.—To move forward at a run from prone position.

vidual rushes from cover to new cover or concealment. You must be able to run at top speed and to drop down and seek cover instantly.

a. Preparing to rise.—Being in the prone position, to prepare to rise, draw in your arms until your hands are opposite your head. Keep your elbows down and away from your body and keep your head down. Avoid careless movements that might betray your preparations to rise (fig. 33).

b. Jumping up.—Raise your body quickly by straightening your arms, shift your weight to your left arm and left leg, throw your right leg forward, and jump off to the right front. Slowness in rising prolongs your exposure.

c. Running forward.—While running, bend your body forward as low as you can and still maintain maximum speed. If armed with a rifle hold it in both hands, left hand near the balance, right hand at the small of the stock, and left shoulder held forward. Run at top speed direct to the next selected point. Normally a single advance in individual rushes should be from 15 to 30 yards when under fire. Slowness in running prolongs your exposure.

d. Dropping down.—Drop to the ground quickly in the most convenient manner, being careful that your legs are not thrown in the air when you fall to the ground. Either of the following methods is easy to execute:

(1) (fig. 34) (*a*) Advance your left foot, turning it so that it points across the front of your body.

(*b*) Drop forward on the outside of your left knee and at the same time extend the rifle grasped in both hands and held vertically, so that the butt strikes the ground at full arm's length directly in front of your left knee.

(*c*) Pivoting on your left knee and the butt of the rifle, roll forward into the firing position, with the rifle at your shoulder, or into the prone position with the rifle on the ground to the front.

(2) While running, jump to a position with your feet on line and 2 to 2½ feet apart. As you jump, shift your rifle so that your right hand grasps it at the balance, and as you land, drop to your knees (fig. 35) Fall forward and catch your weight on your outstretched left hand; stretch your right hand (holding the rifle) forward to assist your left arm in lowering your body (fig. 36). Avoid striking the ground with the muzzle of your rifle. Your legs are stretched to the rear and your left arm is stretched forward. You hit the ground in approximately the correct prone position (par. 30). Slowness in dropping down will prolong your exposure.

■ 33. CREEPING (fig. 37).—*a.* This method of advance is slow and fatiguing. You may have to use it from the prone posi-

FIGURE 34.—Dropping to ground from a run.

FIGURE 35.—Jump and land on both feet. Drop to your knees, rifle low, left hand stretched out to catch your weight.

FIGURE 36.—Lower your body to the ground with your left hand and arm assisted by your right.

tion for short advances, when you are close to the enemy and low cover such as a shallow ditch, low wall, or tree trunk is available.

b. Rest your body on your lower legs, elbows, and forearms. Lift your chest and stomach slightly off the ground. Keep your body low, your head and buttocks down. Keep your knees always behind your buttocks and drag your body forward by alternately pulling with either arm. With practice,

① RIGHT—When creeping, keep your knees behind your buttocks.

② WRONG—Avoid elevating your buttocks.

FIGURE 37.

you will be able to attain fairly rapid movement with very little exposure.

■ 34. CRAWLING (fig. 38).—*a.* This method is slower and more fatiguing than creeping but involves less exposure. You can use it for short advances when you are close to the enemy and have little or no cover.

b. From the prone position bring your hands slowly back beside your head, keeping your elbows down. Draw up either leg and push your body forward with it. When you are not too close to the enemy and when slight cover is available you can crawl faster by rising slightly on your forearms and pushing alternately with each leg.

c. Crawling differs from creeping in that your chest and stomach remain on the ground and you push your body forward instead of pulling it.

d. These methods of movement cannot be acquired by reading, discussion, or thought. They are developed only by constant and intelligent practice. If you have developed the

① When crawling, keep your body and legs flat on the ground and pull with your hands.

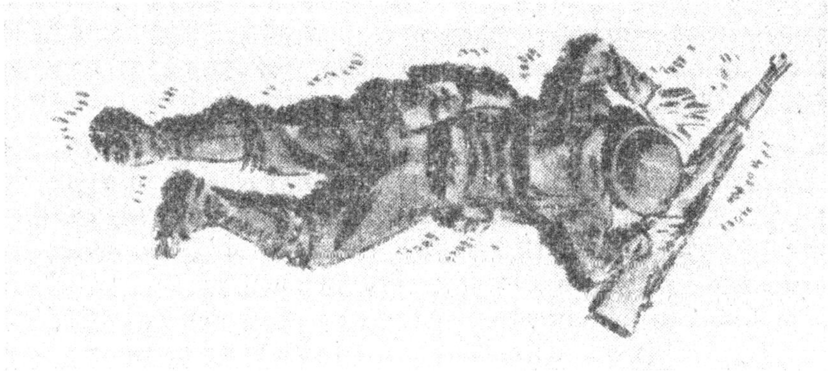

② Draw up either leg and push your body forward with it.

FIGURE 38.

ability to use all of these methods, you will be a valuable member of your unit for *a long time.*

■ 35. COLLECTIVE USE OF COVER.—If you have command of a group you must know how to make full use of the concealment and cover offered by the terrain and the appropriate combat formation to be adopted. It is only with this knowledge that you will be able to avoid unnecessary casualties which might prevent the accomplishment of your mission. All commanders, from the squad leader up, must use the terrain to the best advantage to avoid unnecessary exposure of individuals. If your group is acting as a part of a larger unit you cannot always choose the terrain over which it must advance. You must, however, see that full advantage is taken of the formation used and of all protective features of the terrain over which it moves.

■ 36. METHODS OF ADVANCE.—During the approach march you will conduct your group in the formation designated by your immediate commander. During the actual fire fight you will lead it against the objectives designated by your commander. Usually you will advance from one intermediate objective to another. The actual methods of advance employed by your group in the fire fight are left to your initiative. Terrain and the effectiveness of hostile fire will determine the methods you use. You must take every advantage of the available cover and concealment. Use formations which are irregular and inconspicuous and suited to the terrain. You must coordinate the movement of your group with its own fire and with the fire of adjacent units and supporting weapons. You may have your group advance by infiltration, by using a covered approach, by rushes, or by sideslipping.

■ 37. INFILTRATION.—*a*. This is a suitable method for crossing terrain which affords only intermittent cover for a deployed squad while it is advancing to a forward position. This method involves a minimum of exposure to enemy artillery or small-arms fire, and you may use it very frequently.

b. Indicate the objective or the line on which you wish the squad to re-form. Then direct individuals by name to move forward, directing the scouts to move first. After three or four men have moved forward you should advance yourself. Your second in command should direct the forward movement of the remaining individuals and should be the last to advance.

c. The scouts work forward first to the objective which you indicated. The remaining members of the squad work forward on separated routes to positions approximately in front of their original positions. You must see that they do not bunch or follow in trace over exposed areas. Such exposed areas should be crossed by creeping, crawling, or short individual rushes. You should direct the individuals who follow you to firing positions upon their arrival on the new line and, if fire is to be delivered from the new line, transmit to your men the firing data which you receive from your scouts.

■ 38. COVERED APPROACH.—If the ground affords a covered route, for example a ravine or ditch, you should use it to advance your squad. Avoid long, circuitous routes as they entail delay and loss of contact with units to your right and left. Form your squad in column of files at varying distances and advance it as a unit. You should lead the squad with one or two scouts preceding you, your second in command at the rear. Have the men cross exposed areas along the route by creeping, crawling, or short individual rushes.

■ 39. ADVANCE BY RUSHES.—You will ordinarily resort to rushes under fire only when moving your squad from cover to cover across short stretches of comparatively open terrain or to close with the enemy. It is a very tiring method of advance and should be used only for very short distances. You may advance by rushes of the entire squad, of groups of several men, or by individuals. A period during which the hostile resistance is neutralized by strong concentration of supporting fires favors a rush by a large group. Rushes by smaller groups or by individuals permit those not moving to cover the movement with fire. The rush should be executed at top speed.

■ 40. SIDESLIPPING.—When you are confronted with an exposed fire-swept area in your area of advance you must use your initiative to advance your squad without incurring heavy losses. Where adjacent squads are advancing you may assist them by fire or you may sideslip temporarily to maneuver in their area to knock out, from the flank or rear, resistance in your own area of advance. A squad habitually sideslips to outflank enemy resistance when lack of cover or the strength of enemy fire prevents continued forward movement in its assigned zone of advance. Your platoon leader normally will direct such maneuvers. Adjacent squad leaders must be forewarned of this maneuver, either by prearranged signals or by messages. If you are using this method on your own initiative you are responsible that the adjacent squad leaders are forewarned. It is far better to sideslip and use a nearby covered approach than to risk losses by advancing straight to the front by rushes or even by infiltration. The test of your leadership is your ability to lead your squad to its objective with a minimum of losses.

NOTE.—For additional information on section I see FM 5–20, TF 7–234, and FS 5–3 and 5–10; on section II see FM 5–15, TF 7–35 and 7–234, and FS 5–11; on section III see FM 7–5 and TF 7–234.

CHAPTER 3

SCOUTING AND OBSERVING

■ 41. GENERAL.—*a*. A scout is a soldier whose duty is to reconnoiter and gain information of the terrain or of the whereabouts, movements, and dispositions of the enemy. Since, in modern war, enemy troops may appear anywhere in the theater of operations, it is necessary that you be able to act as a scout no matter what your arm or service.

b. You must have a knowledge of terrain features, map reading, and the ability to determine direction at all times. You should know how to estimate the enemy's strength, the composition of his troops and his actions, and how to report what you have seen by the use of messages and simple sketches and overlays. You must know how to conceal yourself, how to move, alone or in charge of a group, across country to a destination, how to observe, and how to return, without being observed by the enemy.

■ 42. TERRAIN FEATURES.—You must be able to recognize and use the military terms for features of the terrain. Natural features of the terrain are the forms and growths of nature, such as hills, valleys, woods, and streams. Artificial features are the work of man, such as roads, houses, bridges, and railroads (fig. 39).

■ 43. MAP READING.—You should know how to orient a map by compass, by two points, by watch and sun, or by the North Star; understand conventional signs; be able to determine elevation from contours; scale distance on a map; solve simple visibility problems; and find your position upon the map or know your position in relation to other positions on the map or objects on the ground.

■ 44. DETERMINING DIRECTION.—*a*. *With compass*.—You must know how to use the compass for orientation and marching, by day and by night, and to determine direction from a map and apply it to movement on the ground as described in your Soldiers' Handbook (FM 21–100).

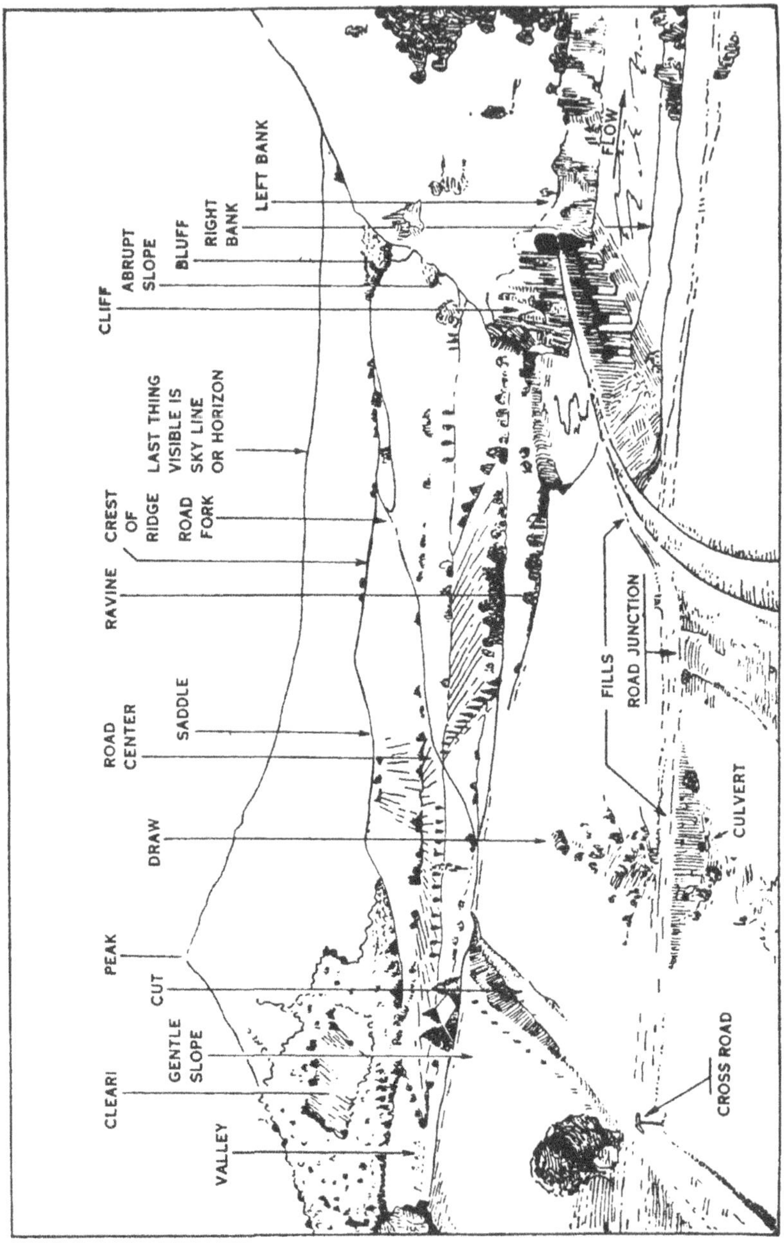

FIGURE 39.—Military terms for terrain features.

b. Without compass.—(1) *General.*—If you are without a compass you should, before going on a mission, study carefully a map of the country to be traversed and fix in your mind the general features, streams, ridges to be crossed, and their relation to the direction you are to take. You should make notes of terrain features and landmarks along your proposed route and rely on your notes for guidance. As you go out on your mission you should look back occasionally to note the relative positions of landmarks, the slope of the ground, and the direction of streams. This will also assist in guiding you on your return. At night, use prominent terrain features outlined against the sky to maintain direction.

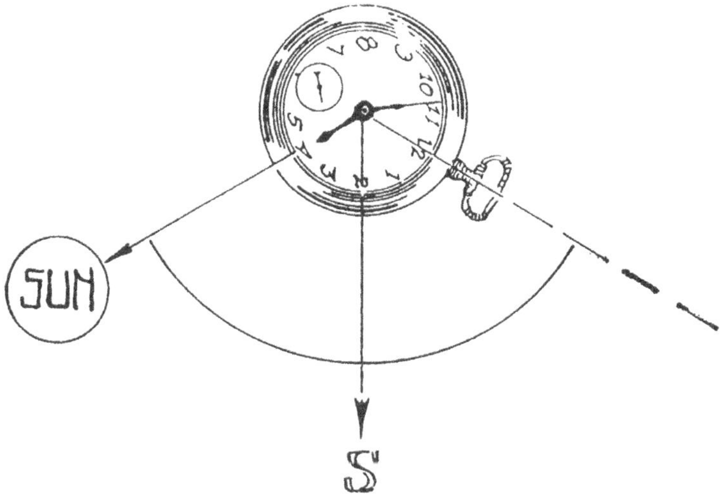

FIG. 40—Method of determining direction by watch and sun.

(2) *By watch and sun.*—Within latitudes of the North Temperate Zone, which include the continental limits of the United States, the following method, correct to within 8°, may be used from 6 AM to 6 PM. Hold your watch horizontally, face up, and point the hour hand at the sun; a line from the center of the dial passing halfway between the hour hand and 12 o'clock (bisecting the smaller angle) will point south. Look along this line and pick out some object in line on the ground (fig. 40).

(3) *By North Star.*—The two stars at the end of the bowl of the Big Dipper, known as the "pointers," indicate at any hour the North Star. The Big Dipper revolves around the

North Star and the pointers always indicate its position (fig. 41).

(4) *When lost.*—Remain calm; halt, try to recall features of the route passed over, and orient yourself by landmarks or the lay of the land.

■ 45. ESTIMATION OF TROOP ORGANIZATION, STRENGTH, AND CONDITION.—*a.* The ability to estimate the organization, strength, and condition of the enemy's troops and get this information back to your commander may result in the proper and early disposition of your unit to defeat the enemy

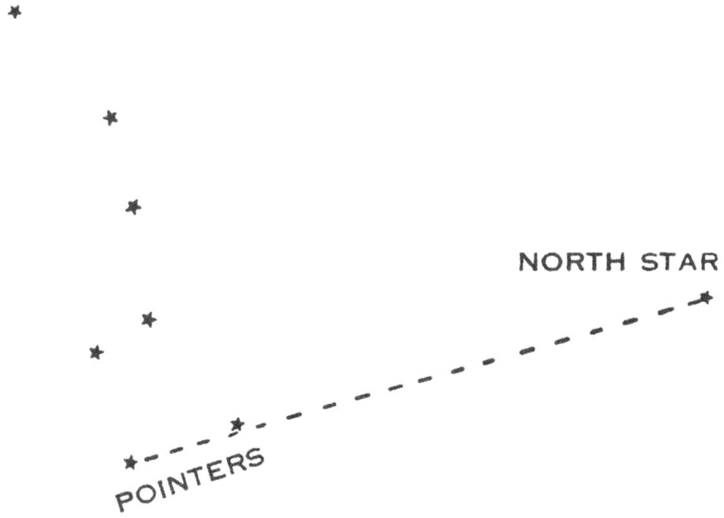

FIGURE 41.—To locate North Star by means of Big Dipper.

force. You should know the organization of the combat units of our own and of the enemy's forces. By observing troops of our Army in camps, on the march, and on maneuvers you will become familiar with their strength and composition. You should note their road space, the front they can cover in deployment, and their appearance under varied conditions. Thus you will learn to estimate the strength of similar enemy units under similar conditions.

b. You may estimate the strength of a column on the march by noting the time required to pass a given point. For example, Infantry in column of threes occupies 0.8 yard per man; therefore, there are 125 infantrymen in such a column 100 yards long. Cavalry in column of fours occupies

1 yard per man; thus there are 100 cavalrymen in a column of fours, 100 yards long. Horse-drawn artillery in single column occupies about 20 yards per gun or caisson. On an average, a point will be passed in 1 minute by 110 infantrymen in column of threes, 110 cavalrymen at a walk in column of fours; and 5 horse-drawn guns or caissons at a walk. The number of vehicles in a motorized or tank column to pass a given point will vary with the speed of the column and the distance between vehicles. To estimate the number of vehicles in a given column, count the number of vehicles passing a given point in 1 minute and multiply this number by the number of minutes it requires the column to pass a point.

c. Lights, fire, smoke, dust, or noise may give information as to the strength, composition, and actions of an enemy force.

d. (1) You may frequently gain information of the enemy by a study of tracks. If you are familiar with the enemy organization and equipment you can, by noting various tracks, make an accurate estimate of the composition of his forces. The following points will aid you in acquiring information from tracks:

(*a*) Different armies and sometimes different organizations in the same army use different kinds of footwear. You should be able to distinguish between the tracks made by various kinds of footwear, friendly and hostile.

(*b*) A large column wears a dry road smooth and flat. Comparatively little dust will be deposited on roadside vegetation after the passage of foot troops, more after mounted troops and trucks, and a great deal if tanks have passed on a dry road.

(*c*) Small wheel tracks may indicate the passage of machine guns, antitank guns, mortars, motorcycles, or other small reconnaissance vehicles.

(*d*) Artillery, tanks, and supply columns make very distinctive tracks. The passage of Diesel-powered tanks and other vehicles can sometimes be detected by the distinctive, persistent odor of Diesel fuel.

(*e*) The condition and morale of troops are shown by the following points: the distance between hourly halts may indicate the rate of march; the ground cleaned up after the halt indicates good discipline; rubbish, packs, rifles, and ammunition scattered about show low morale and poor discipline; tracks leaving a column in the direction of orchards, farmhouses, and wells indicate poor discipline; heaps of stores and materials in good condition left by a retreating force indicate a hasty withdrawal or rout; burned supplies and destroyed materials indicate a more orderly withdrawal.

(*f*) The speed and direction of a vehicle are shown by the following (the side of the road the tracks are on and the side of the road it passes others vary according to custom of the country): a car passing through mud or water will show wet tracks on the side on which it leaves the mud or water; mud and water are scattered more by a swiftly moving car than by one moving slowly; piles of dirt and sand are scattered by a swiftly moving car whereas slowly moving wheels leave deep, smooth tracks; a wheel going over a hole in the ground leaves a deeper mark on the side toward the direction of travel—the greater the speed the deeper this mark will be.

(2) You should practice tracking on soft ground, then proceed to more difficult soil, and finally work on rock ground. In this last stage you should not be discouraged if at first the trail appears blank. Constant practice will bring improvement. In tracking, do not continually look at the ground at your feet. Form the habit of looking forward 20 or 30 paces where scratches on hard ground, bent blades of grass, or broken twigs, which viewed individually mean nothing, form a line and indicate a trail when seen at a distance.

(3) If the tracks which you are following merge into many others and there is a chance of losing them, you should take accurate measurements and attempt to pick them up where they emerge from the confusing ones. At times the tracks will appear entirely lost and you must make a search to relocate them. If there are others with you have them halt before their tracks blot out or are confused with those which you are following. Designate one man to proceed in the original direction while another goes about the spot where you last saw the tracks, in circles of ever increasing size. In making such a search you should put yourself in the enemy's place and figure out what action would be natural under the circumstances.

■ 46. MESSAGES.—*a*. Information which you acquire must be received by your commander in time for him to act upon it. Information may be reported either orally or in writing. An oral message is used when writing is impracticable, when the message contains but one simple idea, or when the danger

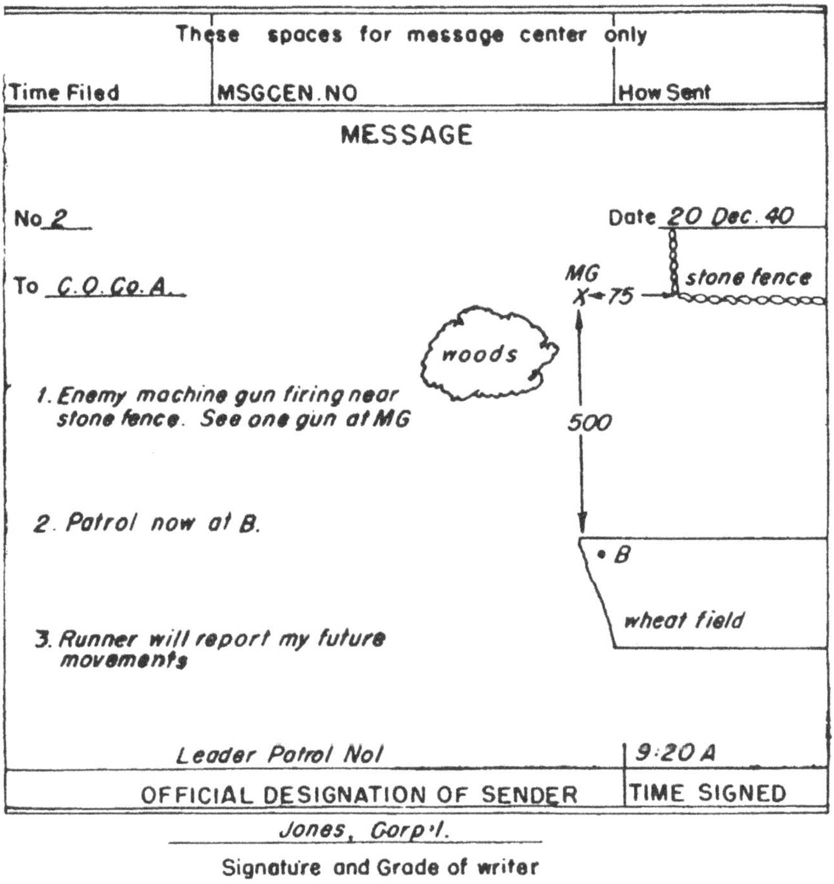

FIGURE 42.—Standard message blank and type message.

of capture makes it unsafe to place the information in writing. An oral message must be clearly understood by the messenger and he should be required to repeat it before starting on its delivery. Written messages should be on the prescribed field message blanks if they are available, and the prescribed form should be followed (fig. 42). You should number your messages in sequence for each mission on which you are sent. If circumstances require you to send a message to any other than your commander, that fact should be stated in the next message to your commander so that he will understand the break in sequence. The date and hour of the occurrence reported should ordinarily be shown. If it is not shown, it is assumed that the event took place at the time shown under "Time signed." The place from which the message was sent should be given. If you are not at a definite point which can be located by map coordinates or is easy to describe, you may describe it by giving its magnetic azimuth from each of two definitely located points, or its azimuth and distance from one known point. At times your position may be better described by means of a simple sketch or overlay. (See FM 21–35.) Such a sketch or overlay may also clarify the other information contained in the message. Such information as the name and position of your organization, which would be of value to the enemy, should not be included in the message.

b. Write your message plainly and state clearly the information which you wish to send. If you have any doubt as to whether an earlier message reached its destination, include a summary of the previous information.

c. Be sure to distinguish clearly between facts and your estimates, beliefs, or opinions. Facts are recorded in a message in unqualified statements; for example, "An enemy machine gun in position at RJ 492." Report other information with a statement as to its source; for example, "A farmer stated that there is an enemy machine gun at RJ 492." Deductions are indicated as estimates, beliefs, or opinions; for example, "Column of horse-drawn artillery, 800 yards in length, moving north, cleared RJ 316 at 8:00 AM." The exact locality where an event occurred or where an object is seen should be given; if not, it is assumed that the event occurred at the place from which the message was sent. After writing a message, you should tell the messenger your intended future action; for example, to continue on your mission, to remain in observation, or to proceed to a designated locality. Your commander will ask the messenger what you are going to do. Thus, if the messenger is captured, there will be nothing in the message to tell the enemy where you are *now*. Read the message carefully after you have written it and, if practicable, have some other person read it in order to check the clarity of its meaning. When signing a message use only your surname and title, as *Smith, Corporal.* Messengers must know where the message is to be delivered and the route to be followed. In friendly territory and close to friendly troops, one messenger is sufficient. In hostile territory, or when it may be necessary to pass through artillery concentrations, two messengers go together or two messages are sent by different routes.

d. Whether information should be sent back at once depends upon its importance and upon whether delay in transmitting the information will detract from its value. If you are in doubt, send the information at once.

■ 47. SKETCHES AND OVERLAYS.—A combat message or reconnaissance report will always be clearer if a sketch or overlay of the area involved is included with it; in some cases a good, clear sketch or overlay will make a written message unnecessary. You should study the methods given in FM 21–35 and be prepared to make simple sketches and overlays.

■ 48. ROUTES.—If a map is available you should study it before starting on your mission. A study of the map will help you to select concealed routes and observation points and to plan your actions in advance. Before starting on a scouting or observing mission you should—

a. Decide where you must go to accomplish your mission.

b. Get a picture in your mind, from your study of the map, of the ground which you must traverse. Note the kind of woods through which you must pass. Open woods offer few obstacles to movement while underbrush may be impenetrable. Brush is generally thicker in valleys and ravines than on summits and ridges. The edge of a swamp or stream bed or a fence line usually offers a concealed route. Note the probable dangerous areas such as crossroads, villages, or high points where hostile observers may be stationed. Consider the effect which the weather of the past few days may have had, particularly if you must cross low ground, creek bottoms, swamps, or streams. Remember that many small features of the terrain not shown on the map will offer you concealment. Select a route which will offer you the most concealment (fig. 43). Determine the compass direction at the start and the reading at each change in direction. Pick out intermediate points from which you can observe along the route. You should consider your return route while studying the map. Frequently it will not

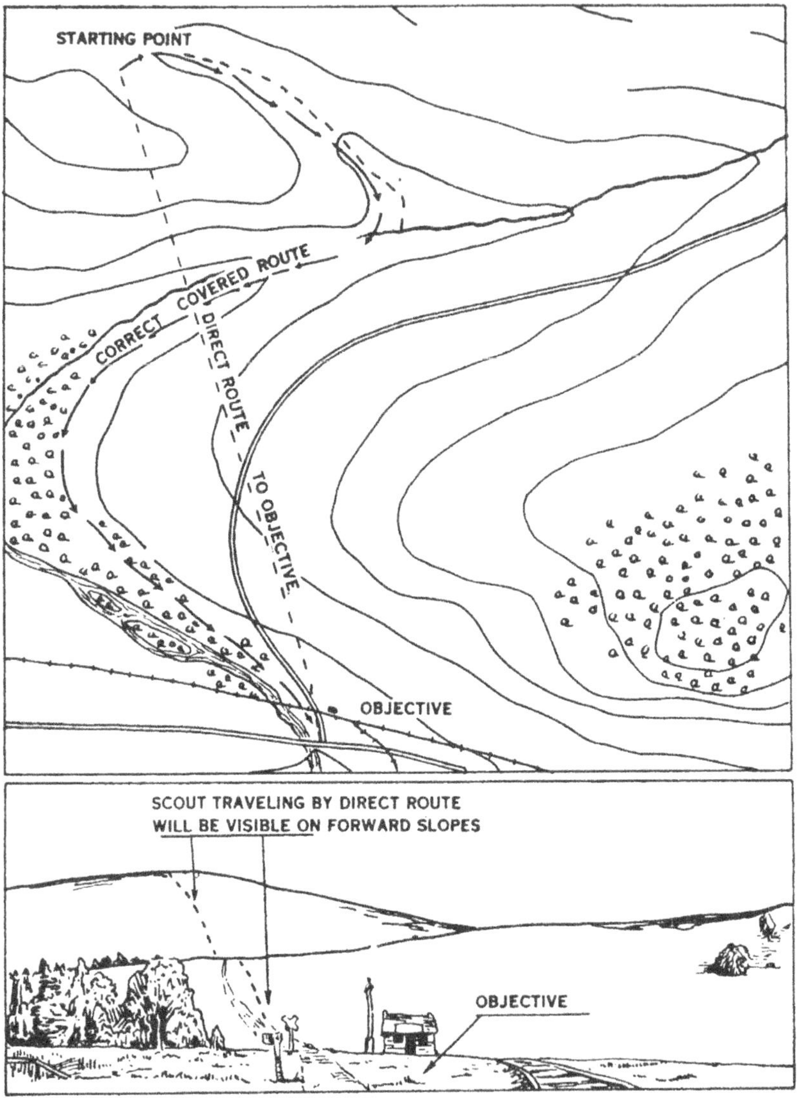

FIGURE 43.—To choose concealed route of advance from map.

be wise to return by the same route and you should always be prepared to return by a different one. Your best guide to your return is your memory of landmarks which you passed on the way out. If you wish to return by the same route you may mark it as you go out by breaking branches, blazing trees, or knotting high grass. Such marks should be made so that you can see them readily on the way back.

■ 49. CONCEALMENT (see ch. 2.).—You must become expert in the use of concealment since your duties may require you to approach very closely or even to penetrate hostile positions without being observed. Observance of the following points will help you remain undiscovered:

a. Be sure that you expose nothing which glistens, that your clothing blends with the background, and that you are carrying no articles which may rattle.

b. Keep in the shade so that you will cast no shadow which might disclose your position.

c. If you are approached by a low-flying airplane, remain motionless, keep your face down and your hands covered, and conceal all maps and papers.

d. If your hair is very dark or very light cover it with grass when not wearing a head covering.

e. Stripe your face with different-colored paints or with mud.

f. A cloak of gunny sacks or sandbags covered with leaves, mud, or clay characteristic of the particular area will aid in concealing you if you must remain in a fixed position for a long time.

g. A small, thin bush in the shadow of a large one is frequently better concealment for observation than the larger one.

■ 50. MOVEMENT BY DAY.—*a.* When you move, move rapidly, silently, and from one concealed position to another. Remain motionless when you are not changing position. Act always as though you are being observed.

b. When you stop, remain motionless. If you are prone, stretch your body out and keep all parts as close to the ground as possible. To observe, move your head slowly and steadily, avoiding all abrupt movements. When you change positions avoid any exposure which would betray your intentions to the enemy. In crossing an open space, spring up, run at top speed with your body bent low, drop, and remain motionless. If a wall or shallow ditch is available, creep behind it, keeping your head and buttocks low. If you are behind a slight rise or very close to the enemy, crawl, keeping all parts of your body close to the ground.

c. In the open, in the presence of the enemy, you should move by rushes from one concealed location to another. Be-

FIGURE 44.—Scout's route of advance showing use of concealment.

fore starting, select your route according to the concealment and cover afforded and the activity of the enemy. You may need to make wide detours around open spaces or areas containing hostile patrols. Your advance seldom will be in a straight line, for you must move along hedges, hollows, woods, ravines, or other concealment (fig. 44).

d. Pick out those places from which the enemy may be observing and move as though you were being observed from each one. From one concealed position pick out your next stopping place. Choose an inconspicuous place which offers good concealment. Avoid the most obvious places of concealment as they are conspicuous and will be the ones most closely observed by the enemy. Before leaving one position observe the next stopping place until you are sure it does not conceal an enemy.

■ 51. Aids in Movement.—The following points will help you move without being observed:

a. You move more freely when lightly equipped than when heavily burdened. Carry only necessities.

b. Don't disturb birds or animals if you can avoid it. When they have been disturbed, remain motionless, for they may have attracted attention to your position.

c. Any incident which diverts attention, such as an airplane fight or sudden bursts of fire, gives you an opportunity to move.

d. If the enemy becomes suspicious of a disturbance at some point, he will watch that point to the exclusion of the rest of the landscape. Take advantage of this by remaining motionless if you are near the point of disturbance or by moving forward if you are some distance away.

e. Fog or even light haze offers concealment to movement.

f. It is best to swim a river or lake at night when you are in the presence of the enemy. If you must do so in the daytime, make a small raft of a few sticks or brush and tufts of grass for concealing your head. In swimming a stream first study objects drifting in the current. Follow the path of objects which approach the other bank so that the current will aid you and your raft will appear natural.

g. If you must move along a beach keep close to the water's edge. The spray and rolling waves will help conceal you from a boat offshore, and there is less danger of your being seen against the skyline.

h. When moving in tall grass or a wheat field while the wind is blowing, move forward when the foliage is moving. In such a field you should change direction frequently as a straight path is readily noticed.

■ 52. OBSERVATION.—*a. Positions.*—(1) *Choice.*—In order to carry out an observation or reconnaissance mission you must generally occupy one or more observation positions. Before

FIGURE 45.—Method of approaching an observing position.

starting, you should study your mission, make your plans, and select observation points either from a map or the ground. When you arrive near a previously selected observation position you should observe closely for 10 or 15 minutes

FIGURE 46.—Correct use of concealment.

to be sure it is not occupied (fig. 45). Then decide upon the exact point from which to observe. Of several equally good positions choose the least prominent.

(2) *Occupation.*—Having chosen the exact spot from which to observe, move to it by a covered route. If it is on a hill, crawl to a place where the skyline is broken. If you observe

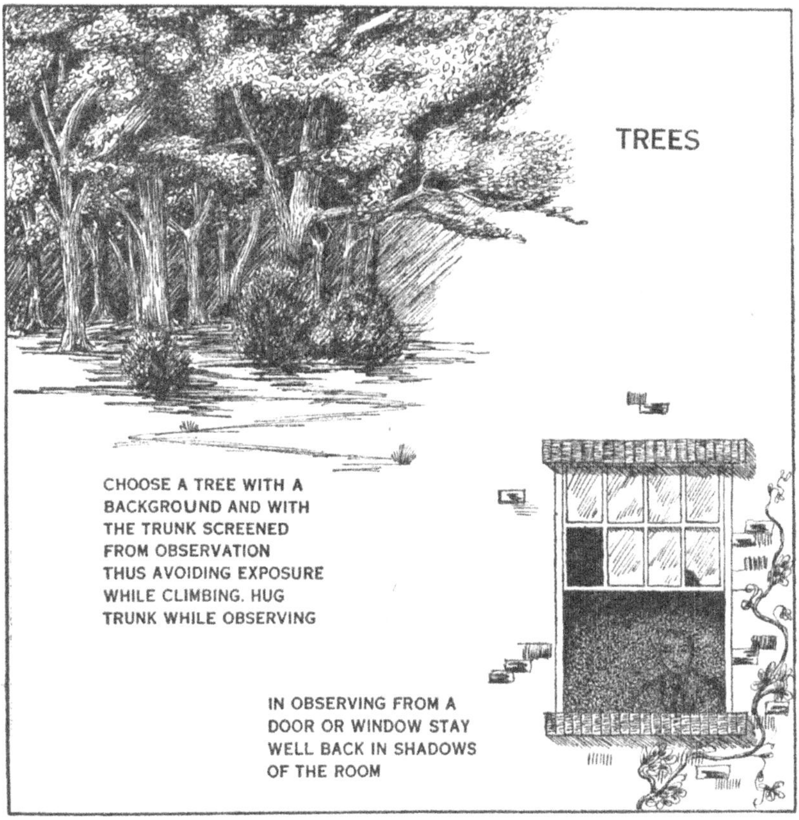

FIGURE 47.—Correct use of concealment; observing positions, showing correct occupation.

from a building, keep back from doors and windows. If you climb a tree, pick one with a background so that you will not be silhouetted against the sky either while climbing or observing. Climb on the side away from the enemy and hug the trunk closely at all times (figs. 46 and 47). While observing, you must avoid unnecessary movement. Leave your position by a route different from that of your approach. When an observation post is to be occupied for several days, care must be taken to avoid making a trail that can be picked up by an observer in an airplane.

b. To search ground.—(1) *Value of training.*—Your value as a scout or observer depends on your ability to see things

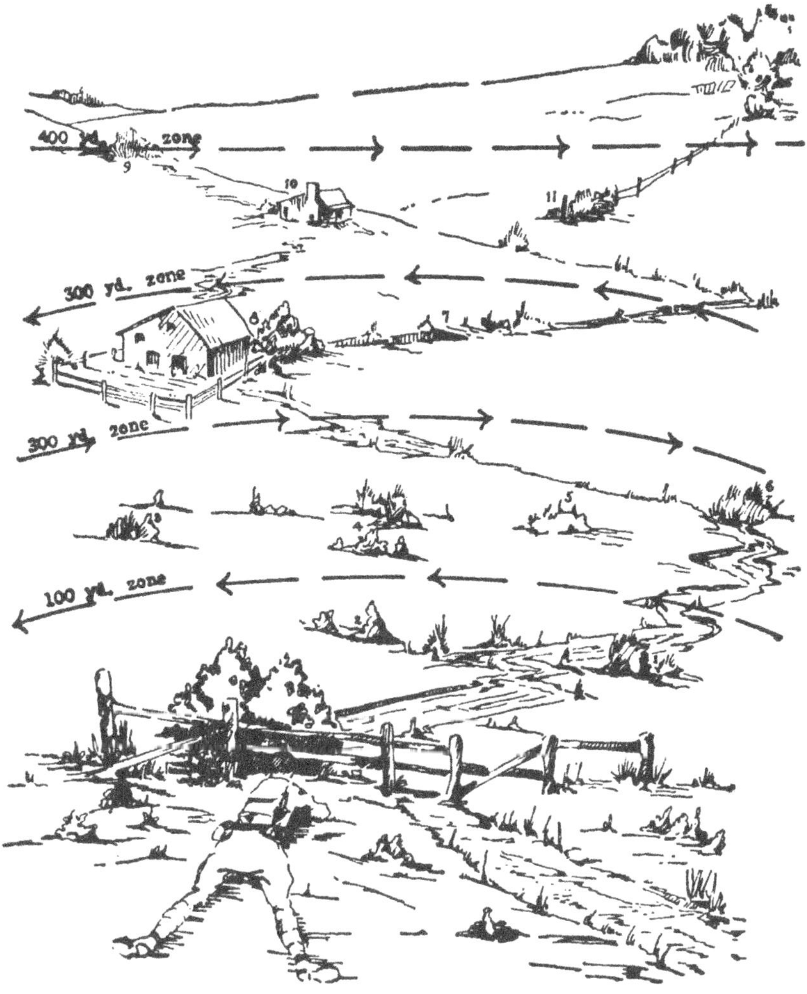

FIGURE 48.—Method of searching ground.

which ordinarily are overlooked. You must learn to observe systematically in order to pick up all objects of military value, both moving and motionless, no matter how indistinct. Painstaking search is often required to locate a position from which fire is coming. The ability to do this is of great value in rifle organizations.

(2) *Method of search.*—Look first at the ground nearest you for your most dangerous enemy will be there. Omit no portion of the dangerous area or place of concealment in your survey. Search a narrow strip close to you from right to left, parallel to your front. Then search from left to right a second strip farther away but overlapping the first. Continue in this manner until the entire field of view is covered (fig. 48). If you think that you see an enemy, look a little to one side of the suspected spot for in this way the eye sometimes responds more quickly to slight movements.

c. To search water.—In looking across a body of water when the sun is shining, the eyes should be shaded from below to shut out the glare from the water.

■ 53. NIGHT SCOUTING.—You must be able to approach hostile positions and to pass through enemy outguards at night. Such missions require different methods from those used during daylight.

■ 54. ROUTES.—*a.* Before starting on a night mission you should study the ground in detail from an observation post, during daylight, if possible, from airplane photographs, and from a map (see par. 48). Make certain of compass directions, prominent points, particularly those which you will be able to see against the sky, and (in a stabilized situation) of the location of gaps in our own wire.

b. Your route should be on low ground and *always off the skyline.* Darkness gives you your concealment so, unless the moon is bright, you should use open ground and should not pass through woods, ditches, ravines, and brush because the noise made moving through them may lead to your discovery (fig. 49).

c. You should always return from a night mission by a route different from that of your advance, when possible, in order to prevent capture by enemy patrols which may wait in ambush for you along the route you took in going out.

■ 55. DIRECTION.—For night work you must understand the use of a luminous compass. It will be of great assistance in following the route you have selected and in reporting information which you have obtained. If you have no compass to use at night you must decide before starting on some means of keeping your direction. Useful means for keeping direction at night are the direction of the wind, stream courses, fences, roads, the edge of woods, stars, and prominent points on the skyline (fig. 50). Notes made from a map may be helpful. The North Star is an excellent reference point and you should

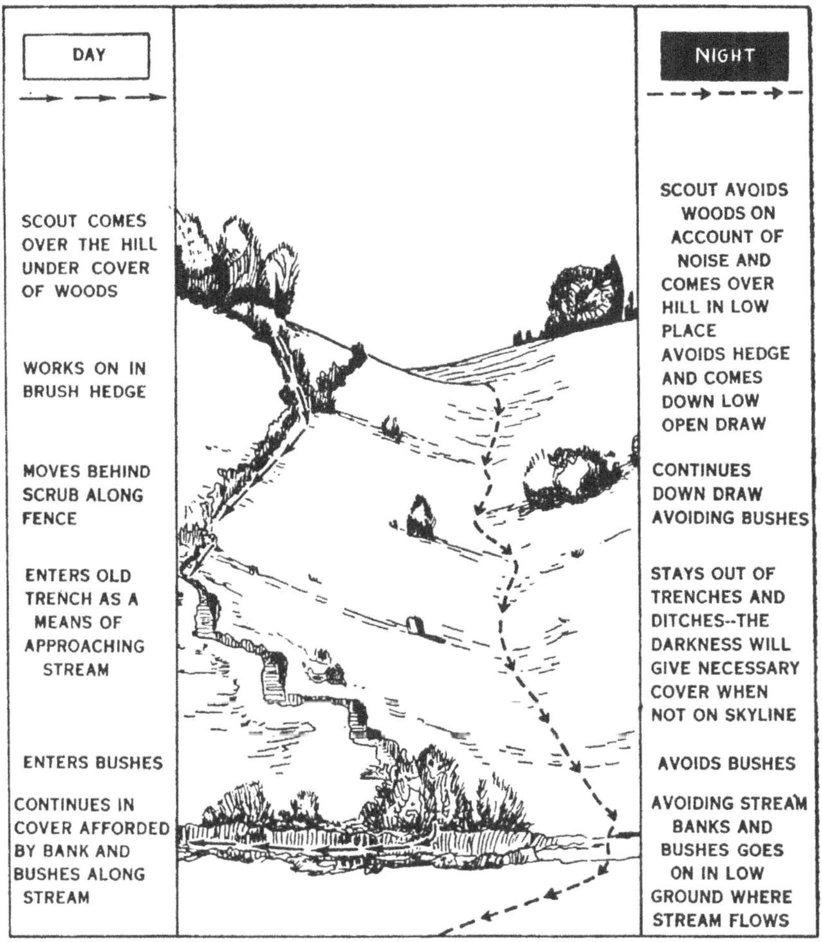

FIGURE 49.—Correct routes over same ground by day and by night.

be able to recognize it. Signal lights may be sent up from your own outguards to help guide you.

■ 56. MOVEMENT.—To accomplish a scouting mission at night you must be able to creep, crawl, and walk silently. Unless the night is very bright, or many flares are being sent up, you may approach by walking crouched over and by creeping to within 100 yards of the enemy position. Beyond this point you must advance by crawling slowly to within 15 or 20 yards of the enemy line; thereafter further movement is possible only by combining careful listening and quiet movement a few inches at a time. You must always keep in mind that at night, when near the enemy, your security depends on silent movement.

■ 57. WALKING (fig. 51).—If you go silently you can make good progress at night because you can walk about unseen.

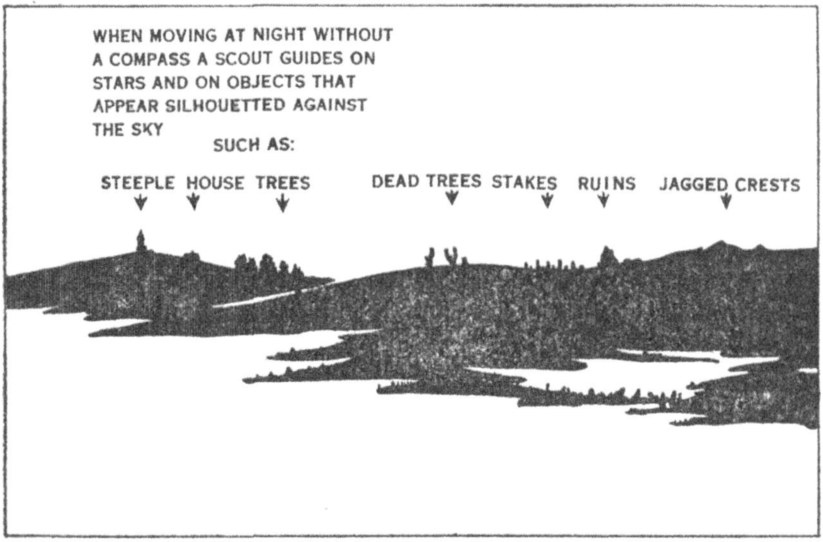

FIGURE 50.—Keeping direction without compass when moving at night.

On soft ground lower your advanced foot, heel first, and then put the ball of your foot on the ground slowly and quietly. On hard ground place the toe first and then gently lower your heel. Through grass raise your foot above the grass before carrying it forward and lower it, heel first, as on soft ground.

■ 58. CRAWLING.—Crawling at night is done the same as by day (see par. 34) except that before each movement you should feel the ground ahead of you to make certain nothing will obstruct your passage or make a noise to reveal your presence. The expert in crawling will be able to get within 5 yards of a sentry without detection, whereas the untrained probably cannot get closer than 50 yards.

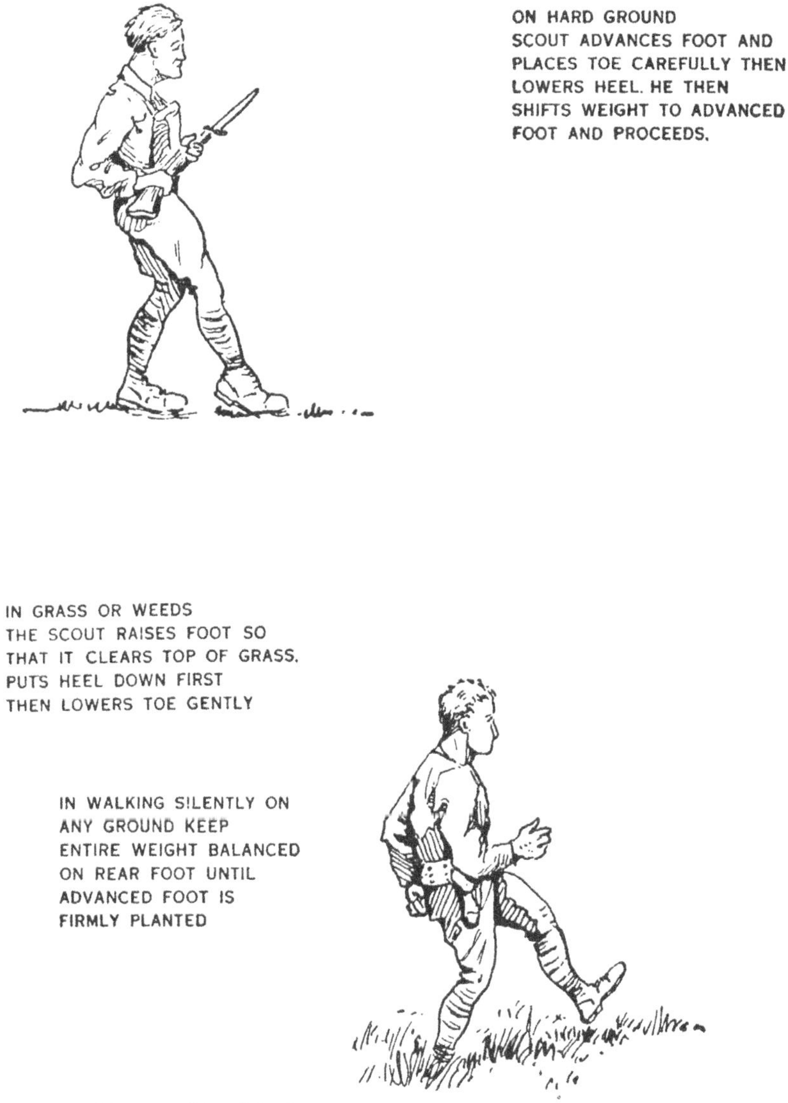

FIGURE 51.—Method of walking silently at night.

■ 59. PASSING OBSTACLES.—Your mission will often require you to pass through and work behind the enemy outguards. To do this you must be able to pass through enemy wire and cross trenches quietly.

a. Passing wire (fig. 52).—(1) To cut a gap in wire requires time and may alarm the enemy. When possible, walk over the low bands and crawl under the high bands. To step over low wire crouch low so that you can see the strands against the sky. Grasp the first strand with one hand and with the

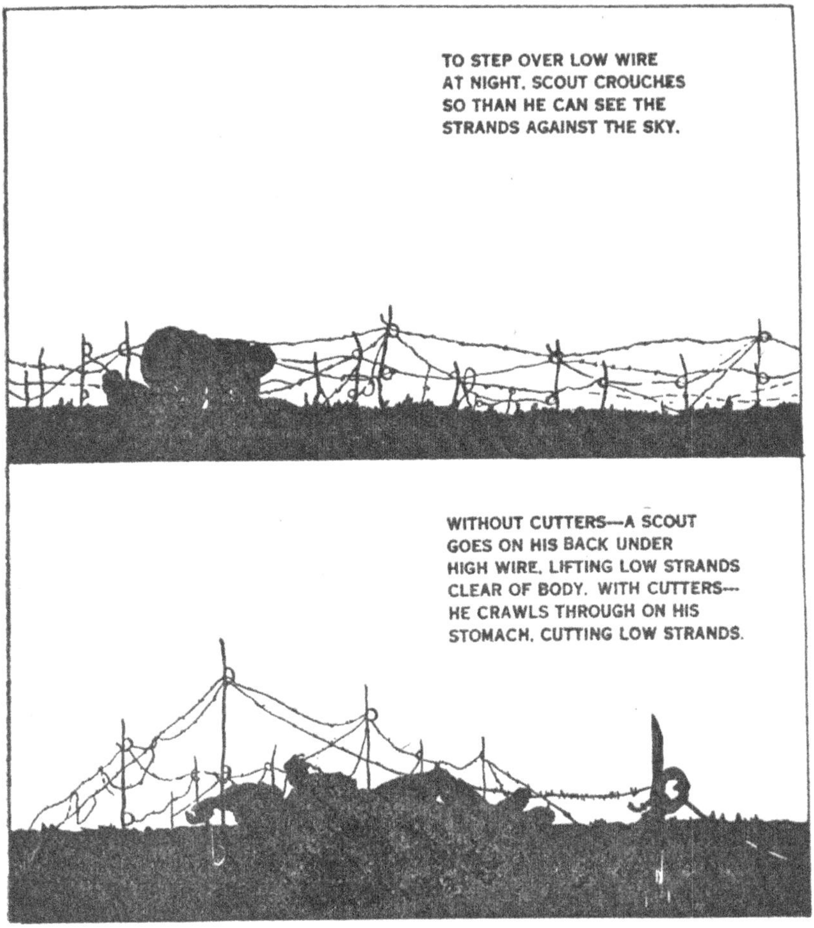

FIGURE 52.—Methods of crossing wir ilently at night.

other reach forward and feel for a clear spot where you can put your feet without stepping on other strands or any objects apt to make a noise. To avoid catching your foot in another strand lift your foot up and over close to the hand which grasps the wire. Go under wire on your back, grasping the lowest strands in your hands and holding them clear of your body while you slide under them.

(2) Barbed wire entanglements frequently contain antipersonnel mines to increase the time and difficulty of passage and give warning to the enemy. Before crossing an entanglement in front of the enemy position, you should examine it as carefully as possible against the skyline in order to locate any large objects against posts or hung in the wire, and any fine wires which might be connected to antipersonnel mines. For a description of antipersonnel mines, precautions to be observed, and methods of dealing with them, see paragraphs 136 to 139, inclusive.

b. Passing antitank mine fields.—Most antitank mines are not detonated by the weight of a man and may be crossed on foot without special precautions unless the field is protected by antipersonnel mines. Since antipersonnel mines used to protect an antitank mine field are mostly operated by trip wires, you are less likely to set them off if you cross the field at a point where the ground is smooth and hard and there is little or no grass or bushes. Your feet should be lifted and set down as though you were passing through high grass.

■ 60. CUTTING WIRE (fig. 53).—At times it will be necessary to cut wire in order to get through it. If you are working alone, cut it near a post. Grasp the wire close to the post and cut it between your hand and the post. You will thus be able to muffle the sound and keep the loose wire in your grasp. Bend back the loose end to form a passage. If another scout is working with you, one should hold the wire firmly close to the wire cutters in order to muffle the sound and prevent the loose ends from flying back while the other cuts. Then each bends back a loose end for a passage. When it is necessary to cut electrically charged wire, special insulated wire cutters and specially trained personnel will be required.

■ 61. CROSSING TRENCHES (fig. 54).—*a.* When you must cross a trench select a point away from its junction with a communicating trench. Before crossing, wait outside for a while and listen, then crawl up to the edge of the trench and look into it. Remove all loose dirt and rocks from the edge. Look at the other side and be sure it is firm and clear of debris.

FIGURE 53.—Methods of cutting wire silently at night.

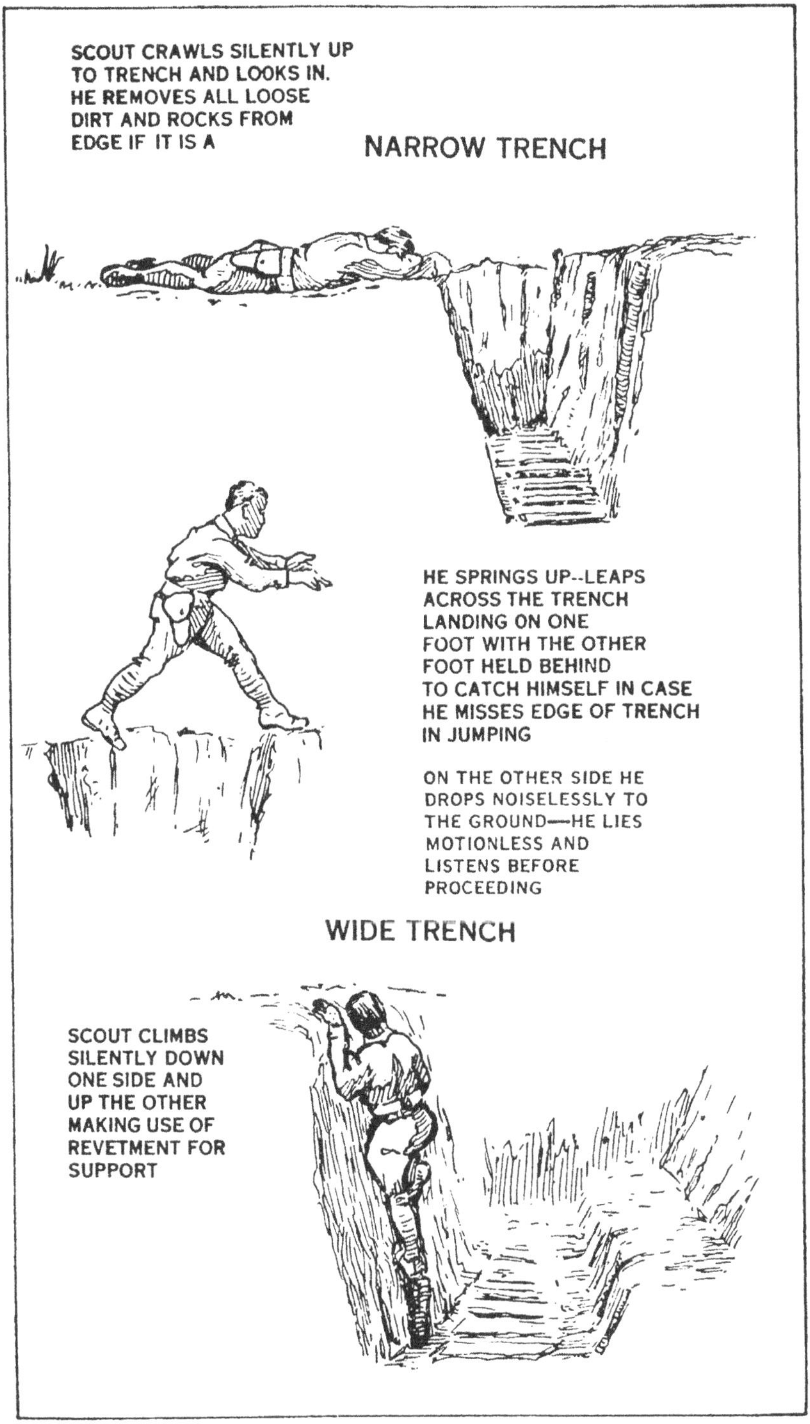

FIGURE 54.—Method of crossing trenches silently at night.

If it is a narrow trench, spring up and jump across, sink quietly to the ground, and remain there for a moment listening before proceeding. If the trench is wide, you must climb silently and slowly down into it and out the other side, using the revetment to assist you.

b. You should not enter an enemy trench, except to cross it, unless you have received specific orders to do so for a definite purpose. Ordinarily, work can be better accomplished from the outside. If you are ordered to enter a trench to accomplish a mission there, you should cross it first and approach the place where your work is to be done from the rear, since sentries pay more attention to sounds in front of them than to those in rear.

■ 62. OBSERVATION AT NIGHT.—*a.* Observation is difficult at night and the information which you acquire will depend largely on your hearing and your ability to interpret sounds. You must cultivate the ability to listen in silence for long periods since, on a night mission, you will have to stop frequently and listen for sounds of the enemy. You must learn to estimate the distance to sounds and be able to distinguish what causes them. Remember sounds travel a greater distance in wet weather than in dry.

b. Constant practice is required to accustom the eye to night work. The distance at which an object can be seen at night is limited. When the eye is close to the ground so that objects appear against the sky, they are more easily seen. Low-powered field glasses will increase your range of visibility at night.

c. You should make notes of the time and nature of the sounds you hear and any evidences of the enemy which you see. If you are observing from a location which can be found on the map you should take the azimuth to the enemy location and estimate the distance to it. Otherwise try to locate it in relation to some prominent point which you can find on the map.

■ 63. AIDS TO NIGHT SCOUTING.—The following points will aid you in carrying out a night mission as a scout or observer:

a. Be sure you have no unnecessary equipment or objects which might make a noise or glisten in the light of a flare. Keys or coins should be removed from your pockets. Don't carry a knife and compass in the same pocket. You should not wear a steel helmet on a night mission because it makes a distinctive noise when it touches barbed wire or other hard substance, it shows an unmistakable outline, and it will prevent crawling with your face to the ground.

b. When you hear the sound of a flare leaving its discharger, drop to the ground before the burst. If you are caught unexpectedly by a bursting flare drop or freeze in position and remain motionless until the light dies down. Never look at a flare; keep your eyes lowered or shaded until the light goes out. The best time to move is just after the light has gone out.

c. When the enemy is using many flares he probably has few patrols out; when he is not employing flares his patrols are likely to be numerous.

d. You can often stop a threatened cough by pressing on your Adam's apple, and a threatened sneeze by pressing upward with your fingers against your nostrils.

e. You can often stop a ringing noise in your head, which interferes with hearing, by yawning.

f. Don't strain your eyes by concentrating too long on one object. If objects blur, lower your eyelids slowly, keep them closed for a few seconds, and then open them slowly.

g. To listen for sounds of men walking, hold your ear close to the ground.

h. Take advantage of any sound, such as shelling, wind rustling, or distant firing, to push forward. Move boldly when firing is going on.

i. If it is necessary to whisper to a comrade, first expel most of the air from your lungs to avoid hissing.

j. Don't chew tobacco, as the sound of spitting is unmistakably human.

k. You should consider all patrols or persons you encounter to be hostile until proved friendly. When you meet someone, crouch low to get the approaching person silhouetted against the sky and at the same time to make yourself an indistinct target if he proves to be an enemy. If you are fired on close to the enemy lines, don't return the fire except to avoid capture. A knife, bayonet, club, or blackjack is much more valuable in individual night work than a rifle or pistol.

l. Be as careful in returning as you were in going out in order to avoid hostile patrols and to keep from being fired on by friendly outguards or sentries.

■ 64. WHAT TO DO IF CAPTURED.—When you are on a scouting mission there is always serious danger of being captured. Before going on a mission be certain that you are not carrying papers which are not necessary for your mission nor any means of identification of your unit. If capture appears unavoidable destroy all maps and papers. You are required to give only your name, your grade, and your serial number to your captors. Refuse to answer all other questions. (See ch. 7.)

NOTE.—For additional information on the subject of this chapter see FM 7-5, 21-25, 21-20, and 21-35; TF 5-12, 7-109, and 7-110; FS 5-1 and 5-2.

CHAPTER 4

ANTIAIRCRAFT PROTECTION

SECTION I

ANTIAIRCRAFT MEASURES

■ 65. HOSTILE AIR OPERATIONS.—*a.* Air operations against which you must take protective measures are observation and reconnaissance, level bombardment and dive-bombing attack, and low-flying attack.

b. The enemy can be expected to use aircraft both day and night to obtain information of the strength, dispositions, and movement of our troops. The location of your unit by a hostile airplane not only endangers the plans of the higher commanders, but is an immediate danger to every individual in the unit since an artillery or air attack may follow.

c. Troops normally will not be the direct target of bombardment by medium or heavy bombardment airplanes. Your unit may, however, be stationed within an area containing installations which are targets for such an attack and for this reason you must be prepared to protect yourself from bombardment. Dive-bombing attacks may be launched against points of strong resistance to the enemy's advance and against concentrations of troops. Under these conditions you may expect aggressive and determined attack from this type of aircraft.

d. Attack by low-flying aircraft is a constant threat to all troops. Such attacks are usually launched with little or no warning. They will normally consist of a combination of machine-gun attack and bombardment with fragmentation or small demolition bombs. Chemicals may also be employed. If chemicals are used, the airplanes may be equipped with chemical bombs or spray apparatus.

■ 66. ANTIAIRCRAFT SECURITY.—*a.* Security from hostile air action is always necessary whether you are resting, in movement, or in combat. When contact with enemy ground forces is remote, you must pay particular attention to protection against hostile aviation. The protective measures taken by the higher commands, pursuit aviation, antiaircraft weapons, night movement, camouflage, etc., will limit but not deny aerial action by the enemy. For this reason you and your unit, when within range of hostile air operations, must take adequate measures to provide protection against air reconnaissance and attack.

b. The protective measures used by small units which provide you with security vary with the situation, the degree of visibility, the cover and concealment offered by the terrain, and the capabilities of enemy aviation. The protective measures are as follows:

(1) *Warning.*—Sentries must be posted to insure that your unit has a timely warning of the approach of hostile airplanes in order that surprise may be eliminated and other protective measures may be taken.

(2) *Concealment.*—Effective concealment makes observation and accurate attack impossible.

(3) *Dispersion.*—Dispersion makes aerial observation difficult and greatly limits the effect of an aerial attack.

(4) *Fire.*—The efficient delivery of small-arms fire on low-flying aircraft and dive bombers will destroy airplanes, will cause the enemy to attack at higher, less effective altitudes, and will reduce the number of such attacks.

(5) *Cover.*—The use of natural and artificial cover limits the effect of an aerial attack.

c. Full use cannot always be taken of all of these measures. For example, the mission of your unit may be such that it will have to move, even though subject to air attacks, in order to arrive at a particular point by a certain time. In such a situation, the effectiveness of air attacks will be limited by making the maximum use of dispersion and advantage will be taken of all available concealment and cover to as great an extent as will not interfere with the accomplishment of your mission. There will, however, never be a situation in which failure to provide warning of an air attack can be justified.

■ 67. WARNING.—*a.* To guard against surprise your unit must detail sufficient antiaircraft lookouts to warn of an impending attack from any direction. You must be prepared to serve in this important capacity. The duties of these lookouts are discussed in section II.

b. When you are operating as a member of a unit and a warning of impending attack is given, you will comply with the orders given by your commander. When you are operating individually and receive warning of an air attack, unless your mission forbids it, you should—

(1) If you are in the open, cease movement.

(2) If on foot in the road, seek cover in an adjacent ditch or concealment in a shadow lining the road. Lie down.

(3) If you are in a vehicle or mounted, clear the center of the road, halt, and dismount. Vehicle brakes should be set. *Do not take cover under the vehicle*. It is the principal target.

(4) If you are in position, bivouac, or billet, seek the nearest cover or concealment and remain motionless.

(5) At night, if you have a light, extinguish it.

(6) Do not look up.

(7) If it develops that the air attack is being made with chemical agents, move upwind or under cover.

■ 68. CONCEALMENT.—*a*. Your best protection, for yourself and your unit, against aerial observation or attack is concealment. The measures for concealment discussed in section II, chapter 2, aim to prevent both ground and air observation.

b. Darkness, of course, provides excellent concealment from air observation if you take proper precautions. Only blackout lights should be used by individuals and vehicles and these should be kept at a minimum and used only in accordance with current instructions. You must be on your guard against hostile aviation using flares. When a flare is dropped, it glows at first for about a second and then burns brightly. When a flare is dropped, stop, look down at the ground and remain motionless until the flare burns out (fig. 55).

c. Be particularly careful that you understand and comply with orders about lighting matches or starting fires in a bivouac area at night. It is very probable that if troops are observed in bivouac an attack will not be delivered at that time, but will be delayed until they are marching out when they will offer a better target.

d. In seeking concealment from air observation or attack, remember that the air observer sees a vast jigsaw of fences, hedges, grass fields, ploughed land, wooded land, clumps of

① WRONG—If you look up or move in the light of a flare you will be seen.

② RIGHT—In the light of a flare, freeze in position.

FIGURE 55.

trees and bushes, houses, isolated and in groups, and the crowded rows of city streets and buildings. He sees it in patches and lines of light and dark rather than in terms of color. Your problem is tying yourself, your fox hole or trench, your gun, or your vehicle into the general pattern that is seen from the air.

■ 69. DISPERSION.—*a*. A dispersed formation lessens the vulnerability of your unit to aerial attack and observation. Dispersion is obtained by increasing the distance and interval between individuals in a unit and between units.

b. Dispersion protects your unit from observation because a closely grouped mass of individuals appears from the air to be a single large object and can be seen from a much greater distance than can the same number of individuals when they are spread out over a considerable area.

c. Dispersion protects your unit from air attack in three ways. It makes it difficult for the attacking force to find your unit and deliver its attack; it makes your unit an unprofitable target for attack, even if discovered, because of the small number of casualties which will result; and, if an attack is launched, it will reduce to a minimum the number of casualties caused by a single bomb or burst of machine-gun fire.

d. If dispersion is to be effective it must be accomplished before an attack is launched. Any attempt to take a dispersed formation when hostile aircraft is within observing distance will not only be ineffective because of the limited time available, but will, by movement, attract attention to your unit and invite attack.

e. You must be constantly on your guard against the tendency to congregate with others in a group. Under stress of danger companionship is very inviting, but it only increases the danger. Your safety will be further increased by staying away from points subject to air attack such as truck parks, ammunition dumps, and artillery positions, except when your duty requires you to be there.

■ 70. SMALL-ARMS FIRE.—*a*. Active defense against hostile air attack is provided by our own Air Force and by antiaircraft units equipped with weapons which have been especially designed for fire against airplanes. These means may be augmented by the use of rifles, automatic rifles, and machine guns which are effective against low-flying aircraft within 600 yards of the weapon. Your unit has been trained to take the necessary measures to use these weapons, when conditions and its mission permit, to defend itself against attack by low-flying airplanes.

b. Properly delivered small-arms fire is a very effective method of countering attacks on troops by low-flying airplanes. When an air attack is impending, the decision as to whether your unit takes cover or fires on the airplanes will be made by your commander. The various conditions on which his decision will be made will vary in each attack. When troops are ordered to deliver the concentrated fire that is possible with the number of weapons that are available to them, they can be certain that the cost to the enemy in airplanes shot down and damaged will be such as to reduce to a minimum the number of future attacks.

c. All low-flying hostile airplanes are suitable targets for small-arms fire. All troops must be fully trained in the delivery of fire against these targets, and imbued with the determination to protect themselves against hostile air attacks by the use of their own small arms. Your training with the weapon with which you are armed will include instruction and practice in its use against airplanes.

d. Low-flying airplanes present very fleeting targets. It is probable that the length of time you will be able to fire at a single airplane within effective small-arms range will not be longer than 6 seconds. All available weapons of your unit must engage these targets promptly in order that an effective concentration of fire may be delivered.

■ 71. TARGET DESIGNATION.—*a.* Low-flying air attacks on ground troops will usually be launched by groups of three airplanes flying in V-shaped formations, by single airplanes, or by airplanes in column. Whatever the formation, it is important that all airplanes be fired upon and that there be a concentration of fire on each. No time will be available after the attack is launched to assign individual airplanes as targets for certain units. Previous planning and training of individuals and units are necessary to assure proper concentration and distribution of fire.

b. Your squad or group will be assigned and trained to fire at a particular airplane as its target for each typical air attack formation. Your fire should be directed at that airplane and no other. If there is an attack by a succession of formations of airplanes the squad leader will direct your squad to cease fire at its airplane in one formation in time to bring your fire on the following formation as it approaches effective range.

■ 72. LEADS.—*a*. When your weapon is fired it takes the projectile a definite time, called the "time of flight," to travel to the target. If you fire at a moving target, such as an airplane, the target will continue to move during the time of flight of the projectile. In order to hit the target, you must aim at the point at which the target will arrive at the end of the time of flight of the projectile (fig. 56). Unless the target is traveling along the path of the projectile, that is, directly at or away from you, you must aim at a point some distance in advance of the target and not directly at it. This distance, called "lead," will depend upon the time of flight of the projectile, which increases with the range to the target, and upon the direction of movement and speed of the target.

b. The lead is designated as a certain number of target lengths. By a target length is meant the length of the target as it appears to you at the time of firing. This apparent length of the target will vary according to the angle at which the airplane is approaching or moving away from you. Aim your weapon the designated number of target lengths in advance of the airplane and along its flight. Remember that the unit of measure, that is, a target length, is the length of the target *as you see it*.

c. Suitable leads for fire on airplanes flying at a speed of 300 miles per hour are as follows:

(1) For all targets diving directly at you or climbing directly away from you, or any target not more than 100 feet away from you, use no lead. Aim directly at the target.

(2) For all other targets, if you are armed with a machine gun or rifle, take a lead of six target lengths.

d. The lead must be changed proportionately for airplanes flying at speeds other than 300 miles per hour.

■ 73. DELIVERY OF FIRE.—*a*. *Range*.—The maximum effective range for small-arms fire at airplanes is about 600 yards.

When you are ordered to fire at airplanes you should assume the firing position as soon as possible after you receive warning of the approach of hostile airplanes, and start tracking the target until it comes within range. You must estimate the range at which you open fire. A good general rule is to open fire only when you can see such details of the airplane construction as wheels, rudder, wing struts, or tail skid.

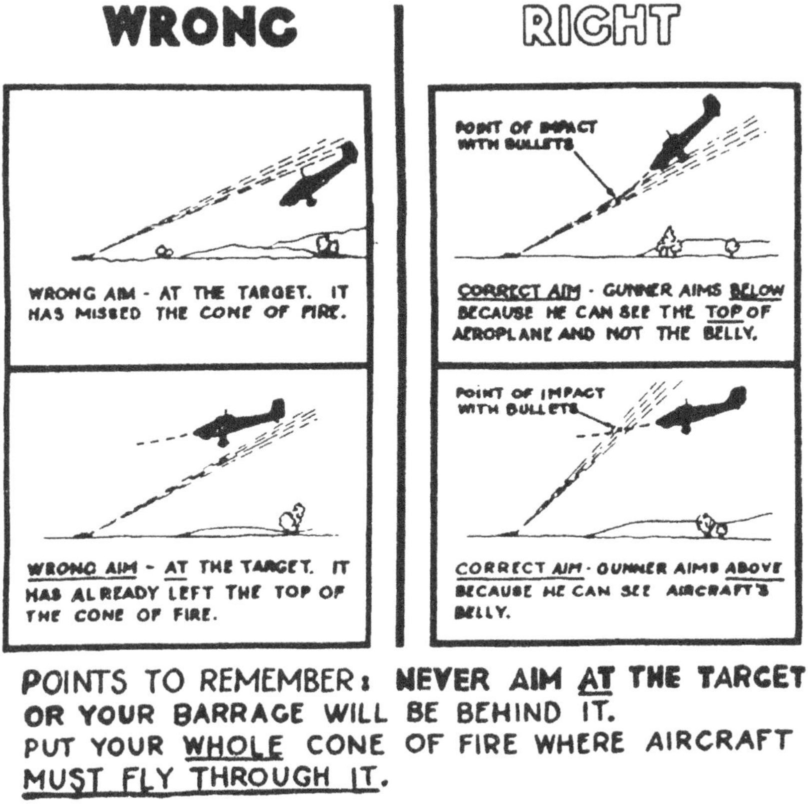

(1) How to aim.

FIGURE 56.—How to hit back at dive bombers.

b. Rate.—Do not sacrifice accuracy of fire for speed. You must aim each shot and squeeze the trigger exactly as in target practice. If you aim carelessly or jerk the trigger you are just wasting ammunition. Three or four shots are probably the most that you can fire properly at a single attacking airplane. Even a single shot, accurately aimed, is better than a clip fired at random in the air.

c. Sight.—Do not try to use the peep sight in firing at airplanes. You cannot use the peep sight and see the target too, which you must do to apply the required lead. If your weapon has a battle sight use it. With other weapons you aim by sighting over the top of the rear sight and the front sight in a manner similar to that used in firing a shotgun.

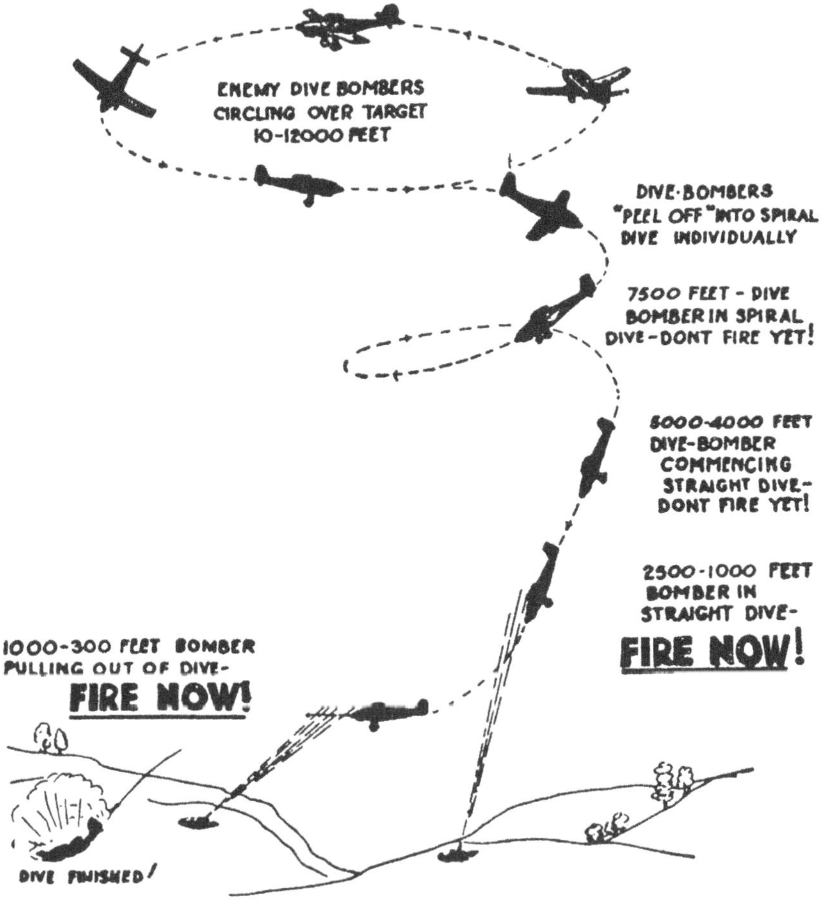

When to fire.

FIGURE 56.—How to hit back at dive bombers—Continued.

■ 74. EFFECT OF FIRE.—*a.* Small-arms fire is capable of various degrees of damage to an airplane. Hits upon important working parts of the engine may stop it immediately. Hits upon the propeller may throw it out of balance with consequent destruction of the engine. Hits upon the control elements may throw the airplane out of control and cause it to crash; even a momentary lack of control at low altitudes is apt to accomplish this result. Hits upon bombs in the racks may detonate them and wreck the airplane. The pilot, even when partially protected by armor, is especially vulnerable. If the airplane is not brought down immediately, it may still be damaged to such an extent that it will be forced to land before reaching its own territory. Hits of any kind require varying degrees of repair, and keep the airplane out of the air for a period of time even though they do not result in its destruction.

b. Small arms have been proved effective, when fired by trained men, against low-flying airplanes. This fire will cause substantial losses to hostile aircraft and will cause them to maintain higher altitudes, from which they are less dangerous to you, and to reduce the number of attacks.

■ 75. RESTRICTIONS ON ANTIAIRCRAFT FIRE.—*a*. Special instructions will regulate the opening of antiaircraft fire. If you are armed with a rifle or automatic rifle you should always be prepared to fire on airplanes, within effective range, unless orders are issued to the contrary. Fire will be withheld when the opening of fire would disclose positions or concentrations of troops where secrecy is desired, as, for example, when you are in a defensive position in the close vicinity of the main line of resistance, or an assembly area from which a surprise attack is to be launched. When positions are known to have been located by the enemy, fire is normally opened on all airplanes flying low over that area.

b. To avoid casualties among our own forces certain restrictions may be placed on antiaircraft fire. The minimum elevation for firing of small arms over or toward our own troops may be designated. Fire should be prohibited at airplanes engaged in aerial combat. Individuals may be prohibited from firing except when taking part in the supervised fire of a squad or other unit.

■ 76. COVER.—*a*. There will be occasions when you will be unable to fire at airplanes, either because of restrictions (par. 75) or because their altitude puts them beyond effective range of your weapons. Proper use of cover will give you excellent protection against the weapons of an airplane. The methods of providing cover have been discussed in chapter 2. The use of all available cover will greatly limit the effect of attack by low-flying airplanes. Cover is necessary against both machine-gun fire and against fragmentation and demolition bombs.

b. The modern light bombardment airplane is capable of delivering a tremendous volume of machine-gun fire at ground troops. Such airplanes usually open fire at about 1,000 yards with the airplane leveled off so as to provide grazing fire. The fire from a single airplane will cover an area about 50 yards wide. Rough calculations show that if an airplane attacking at a speed of about 300 miles an hour is firing eight machine guns, about one shot will hit in each 50 square yards of level ground covered by the fire, which is a large area compared to that covered by one man. The striking points of the bullets, however, represent only a small part of the danger area since the fire is grazing fire and therefore a bullet will hit an object above the level of the ground a considerable distance ahead of the point at which it will hit the ground; and after hitting the ground, the ricochet will still be dangerous. If you are in a prone position under attack by an airplane flying at an altitude of 100 feet, it is about four times as difficult to hit you as when you are standing; and if you are in a ditch, shellhole, or other depression, there is very little chance of being hit at all. As the altitude of the attack increases, which it will as a result of small-arms antiaircraft fire, the grazing effect of the bullets is reduced and the chances of hits on exposed personnel are greatly lessened.

c. In addition to attacking with machine-gun fire, the attacking airplane may drop bombs, usually of the fragmentation type. When fragmentation bombs burst they throw fragments out to a distance of some 150 feet in all directions from the point of burst. If you are standing you have about one chance in six of being hit by a fragment of a bomb bursting within 150 feet of you. If you are flat on the ground your chances of being hit are reduced to one in thirty-four, and if you are in a depression there is practically no chance of your being hit by anything other than a direct hit by the bomb.

d. You can see from the above that any cover, no matter how slight, will afford you protection, and that if you are standing you will be in great danger of being hit. When an air attack is launched seek the closest cover; any ditch, shell hole, depression, rock, or tree will be satisfactory. If no immediate cover is available you will gain considerable protection if you lie flat on the ground. Never run for distant cover since in running you become a perfect target for the airplane and in addition you will attract attention to your unit which may not have been observed.

e. When you are in a position, a trench, fox hole, sandbag parapet, shell hole, or natural depression in the ground will provide adequate protection against anything but a direct hit. You should always provide such cover at the earliest possible moment. (See par. 19.)

■ 77. Chemical Attacks.—Hostile airplanes may attack ground troops by using chemical bombs or chemical spray.

a. Chemical bombs will usually contain blistering gases (vesicants) or white phosphorus. The explosion of a 30-pound chemical bomb upon impact throws its contents over an area approximately 40 yards in diameter, about the same as that covered by a chemical shell. Bombs of 100 pounds or larger may be used to discharge either persistent or nonpersistent chemicals. The action of agents thus released is similar in principle to the action of agents discharged by chemical projector shell. (See ch. 6.)

b. The agents most likely to be used for spray attacks are mustard gas, lewisite, and any type of liquid smoke. Airplanes flying at altitudes between 50 and 1,000 feet can lay a belt of persistent gas in an effective concentration. The width and length of the area covered depend upon such factors as the altitude, course, and speed of the airplane, and the direction and speed of the wind. The drops of chemical will be larger and more effective on the up wind side than the down wind.

■ 78. Protection Against Chemical Attack by Airplanes.—In general, you should take the same measures of protection against chemicals delivered from the air as against those delivered from the ground (see ch. 6). The falling spray itself is an added danger in the spray attack. You must protect yourself against the falling spray by getting out of its way, taking cover from it, or wearing impermeable protective clothing. Avoid looking up as drops of the chemical may cause blindness if they fall in the eyes. Unless you are at a considerable distance down wind from the attacking airplane, you should move into the wind to get out of the spray. Heavy woods will afford you some protection from the falling spray, but you should vacate a wooded area as soon as the spray has stopped falling as the area will be contaminated. If drops of the chemical fall on clothing, prompt action must be taken to remove the clothing to avoid serious burns.

SECTION II

ANTIAIRCRAFT LOOKOUTS

■ 79. GENERAL.—*a*. As long as the enemy has any aircraft that will fly, all troops within range of hostile air operations must expect to be attacked or observed from the air at any time. Continuous security against such hostile air operations must be provided whether troops are resting, marching, or fighting.

b. The steps taken by ground troops for protection against air operations vary with the situation, the degree of visibility, and the terrain (par. 66). The most important protective measure taken is always that of providing timely warning of the approach of enemy aircraft. The more efficient the warning system, the better chance do troops have of protecting themselves against the enemy.

■ 80. NECESSITY FOR ANTIAIRCRAFT LOOKOUTS.—If the task of warning a unit of the approach of enemy aircraft is left to the members of the unit as a whole, one of two things will happen. The attention of all of the members of the unit will stray from the duties they are performing at each sound of an airplane, probably with disastrous results both as to the accomplishment of their mission and their safety from other dangers; or they will become so seriously engaged with their other tasks as to fail to note the approach of hostile aircraft in sufficient time to be able to protect themselves from an air attack. To avoid both of these dangers, individuals are detailed to act as antiaircraft lookouts for their units, thus permitting the other members of the units to concentrate all their efforts on their immediate duties until such time as the lookouts warn that the previously planned protective measures must be taken.

■ 81. Detail of Lookouts.—Every ground unit details antiaircraft lookouts who are on duty at all times. You must be prepared to carry out this important duty in your unit. You may remain at one post, march to the front, rear, or abreast of your unit, or be moved from one post to another by a motor vehicle to protect your unit during movement. At each post you will usually work with another lookout and relieve each other at intervals of not more than 15 minutes. You should be equipped with field glasses, sun glasses, and some means of sounding the alarm.

■ 82. Duties of a Lookout.—*a*. Your mission as an antiaircraft lookout is to warn your unit of the approach of enemy aircraft so that it has sufficient time to protect itself by active and passive means. (See par. 66.) A study of this mission indicates that there are four separate duties which you must perform:

(1) You must be constantly on the alert to hear and see all aircraft in your vicinity.

(2) You must determine whether aircraft within sight or sound are friendly or enemy (see par. 87).

(3) You must determine whether enemy aircraft are close enough and their behavior is such as to threaten your unit with imminent danger from bombs, machine guns, chemicals, or observation.

(4) If danger *is* imminent you must sound the alarm.

b. Your task as an antiaircraft lookout is one of great responsibility. On you depend the lives of many men. If you are slow in identifying an enemy airplane, or in sounding the alarm, the result may be disastrous. If you sound the alarm unnecessarily either on the approach of friendly airplanes, or when hostile airplanes are moving in such a direction as not to threaten your unit, important duties of the unit will be interrupted and your comrades will tend to disregard your later warnings.

■ 83. Training for Lookout Duty.—*a*. Keen eyes and sensitive ears, while necessary qualities, do not in themselves qualify you to function as an antiaircraft lookout. Your unit commander will see that you have thorough, systematic, and continuous training. Since any individual in your unit may be detailed as a lookout, such training must be provided for all the members. Additional specialized training will be provided for those members of your unit who will most frequently function as lookouts. Such individuals should have better than average intelligence and be gifted with good memories, a certain amount of analytical ability, and the ability to make quick decisions.

b. Your training as an antiaircraft lookout will consist of instruction in the following subjects:

Selection of lookout post.
Initial pick-up of airplanes.
Use of field glasses.
Identification of airplanes.
When and how to warn the unit.

c. By far the most difficult subject is that of the identification of airplanes. The rapid movements and different angles at which you see airplanes in flight make positive identification an extremely difficult task for any but thoroughly trained observers. You will get effective training only by the employment of a definite and logical system of identification methods.

■ 84. SELECTION OF ANTIAIRCRAFT LOOKOUT POST.—*a*. You must have a post from which you have a clear view covering the entire area over which you are charged with observing. If you are protecting your unit when it is not moving, a single post may provide an all-around view. If your post is at some distance from your unit you will be provided with telephone or radio communication for giving the alarm. If one post does not have an all-around view, several other posts may be organized and you will be charged with observation of a certain area only. On the march you may be posted at the head, tail, or on the flank of the column, and, in a long column, you may have a post somewhere within the column. If you are in a flank post you may have to march abreast of the column or may be transported by a motor vehicle from point to point along the march route.

b. The necessity for freedom from noise interference at your post is nearly as important as a good field of view. You will often hear airplanes before you see them, particularly when there are low clouds or during bad weather. Your post should therefore be located away from traffic, talking, and other sources of noise, if possible.

c. When practicable you should have a sandbag or other inclosure, not only as a protection against bomb fragments but as an aid to your hearing in windy or stormy weather. Wind whistling by your ear will reduce your hearing range materially, while if you are sheltered in the lee of sandbags, or some other object which will shelter you from the wind your hearing is scarcely diminished. When using any sort of wind protection you should observe while your assistant lookout is listening, alternating these duties at approximately 15-minute intervals.

■ 85. INITIAL PICK-UP OF AIRPLANES.—*a*. Because of the high speed of aircraft it is essential that you be always on the alert to observe the first indication of approach. You must observe in every direction with special attention to that from which attacks may be expected. The direction of the sun, or of hills, woods, valleys, stream beds, or other concealment which might screen low-flying attacks until they are close to your unit, are particularly dangerous. You should be equipped with sun glasses to protect your eyes when observing in the direction of the sun.

b. The most frequent, the best, and in many cases the only evidence of enemy airplanes in your vicinity is antiaircraft and automatic weapon fire of other units. When you see tracers, or shells bursting in the air it is evident that hostile airplanes are in the vicinity. The fact that antiaircraft guns are not firing at an airplane which you see or hear may indicate that it is a friendly airplane, but you should not take it as conclusive proof, since orders may have been issued forbidding antiaircraft fire. In thick, cloudy, or misty weather the sound of gunfire, shell burst, or bomb burst may be the only indication available to you of the approach of hostile aircraft. You must develop your ability to judge how far away gunfire is, since it can be heard from a great distance and it will not be necessary to warn your unit unless the airplanes are approaching it.

c. At night the action of antiaircraft searchlights in moving steadily across the sky is generally an indication of the approach of enemy airplanes.

d. You will frequently locate airplanes by sound. Remember that sound passes through the air at a comparatively low speed and that the airplane may be some distance in advance of the point from which the sound appears to come. A continued irregular whining of diving and zooming airplanes high in the air is a sure sign of an air battle and, if it is close overhead, may indicate the possibility of a rain of bombs, since bombers frequently unload to increase their chance of escape. Similar sounds close to the ground, accompanied by sounds of the bursting of bombs and automatic weapon fire, indicate an attack by low-flying airplanes on nearby units.

e. Vapor trails in the air indicate the passage of an airplane at extreme altitudes and you should observe them closely.

f. The action of friendly pursuit airplanes may indicate the presence of hostile aircraft. The location of the enemy will be indicated by the direction of flight of the friendly fighters as they move toward him.

g. When adjacent units are within your sight or hearing your first indication of the approach of a low-flying aircraft attack may be the sound of an alarm in such a unit or the sight of members of the unit taking cover.

h. You must not become so engaged with the attempt to identify some distant airplane that you forget to observe for closer and more dangerous threats. Teamwork between you and the other lookout stationed with you, in which one of you maintains observation throughout your sector while the other attempts to identify an airplane, will eliminate this danger.

■ 86. Use of Field Glasses.—Ordinary eyesight is not sufficient to identify an airplane at distances sufficiently great to provide the warning necessary to protect your unit against air attack. For this reason you will be equipped with field glasses which you should habitually use in observing aircraft. With field glasses you will be able to identify airplanes, in good weather, at distances up to 5 miles. You should memorize the setting of the field glasses which will adjust them to the distance between the pupils of your eyes and will focus each eye piece. Practice in the proper use of field glasses will accustom you to their use.

■ 87. Identification of Aircraft.—*a.* You must be able to detect quickly and analyze rapidly every possible indication of the identity of aircraft, whether observed singly or in groups. The first requirement is the ability to distinguish rapidly between hostile and friendly airplanes. Having identified an airplane as hostile you should be able to determine its type to enable you to predict its possible action and the resulting danger to your unit. This means you must be familiar with the combat characteristics of the various types of hostile aircraft. (See par. 65.)

b. The only way to acquire the ability to identify aircraft is by constant study of the characteristics of the different types, and by practice in actually observing and listening to aircraft under all conditions of flight. In order to recognize any one of many types, at a distance and from any angle,

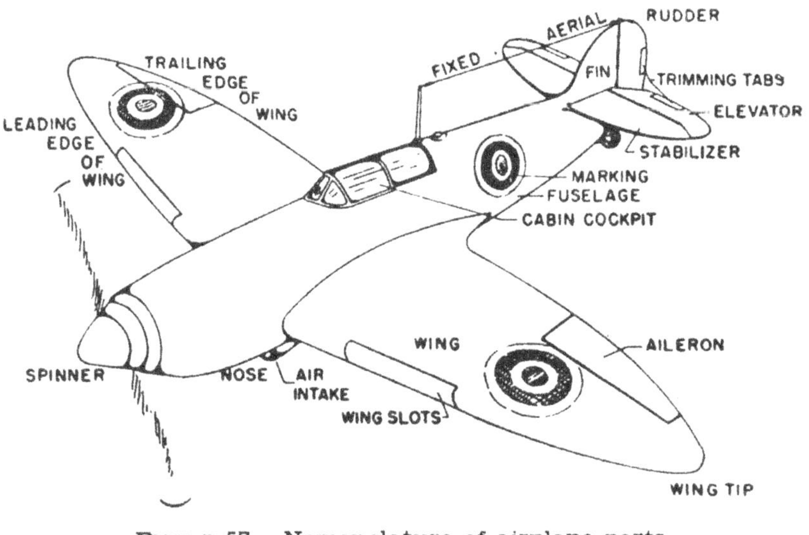

FIGURE 57.—Nomenclature of airplane parts.

it is not sufficient to have seen a number of photographs or silhouettes and to have read a general description of the characteristics of the airplanes illustrated. You must first learn what to look for in an airplane and then to analyze each airplane systematically (figs. 57, 58, and 59). With proper training and sufficient practice identification will become automatic and instinctive.

c. The indications by which you can identify an airplane may be grouped in three general classes:

(1) *Characteristic sounds.*—Generally sounds will permit you to identify only the type of airplane, that is, pursuit, light bombardment, dive bomber, etc. With a sensitive ear

and considerable experience you may learn to distinguish between friendly and hostile types.

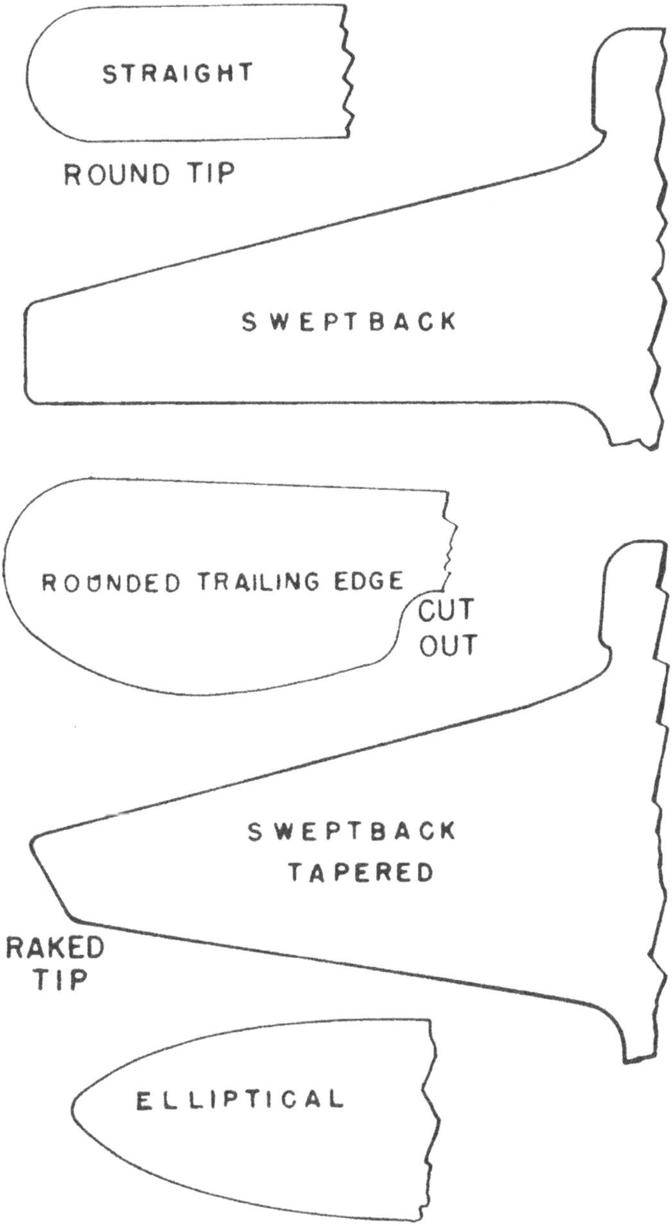

FIGURE 58.—Wing shapes.

(2) *Characteristic methods of operations and maneuver.*— These will indicate to you the types of airplanes and may

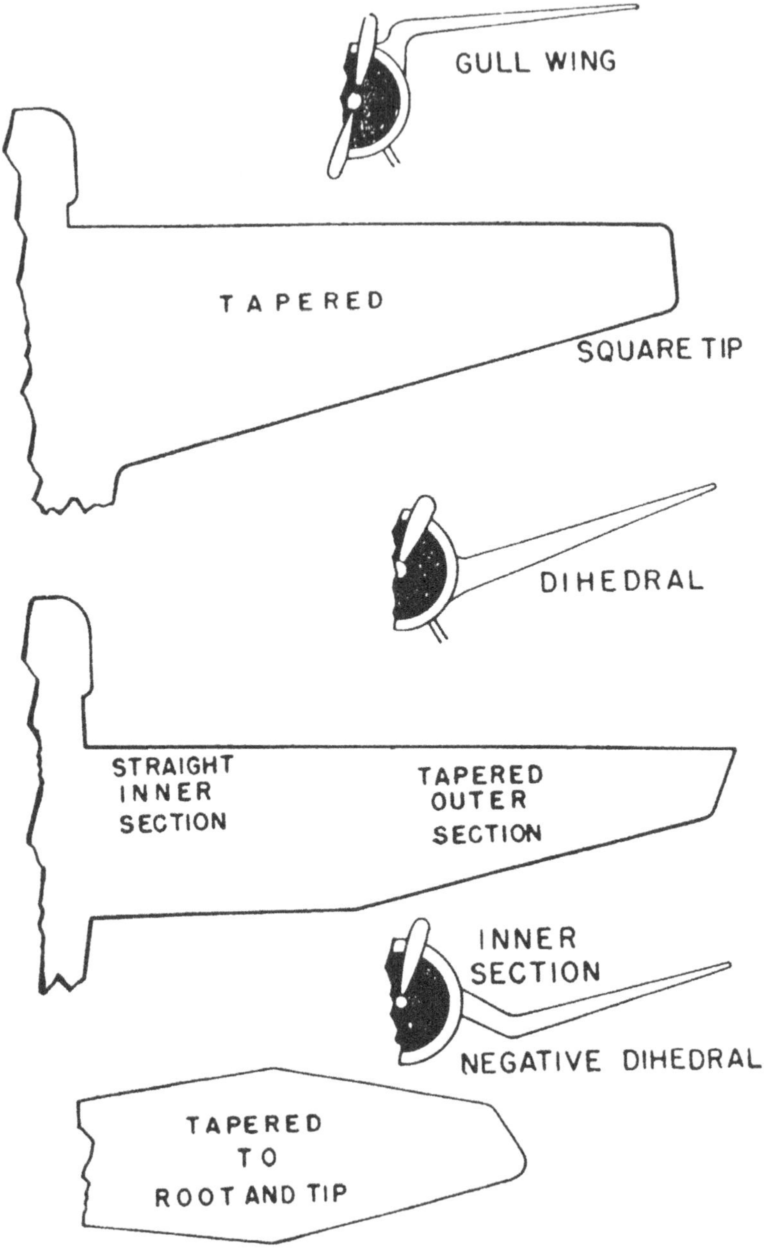

FIGURE 58.—Wing shapes—Continued.

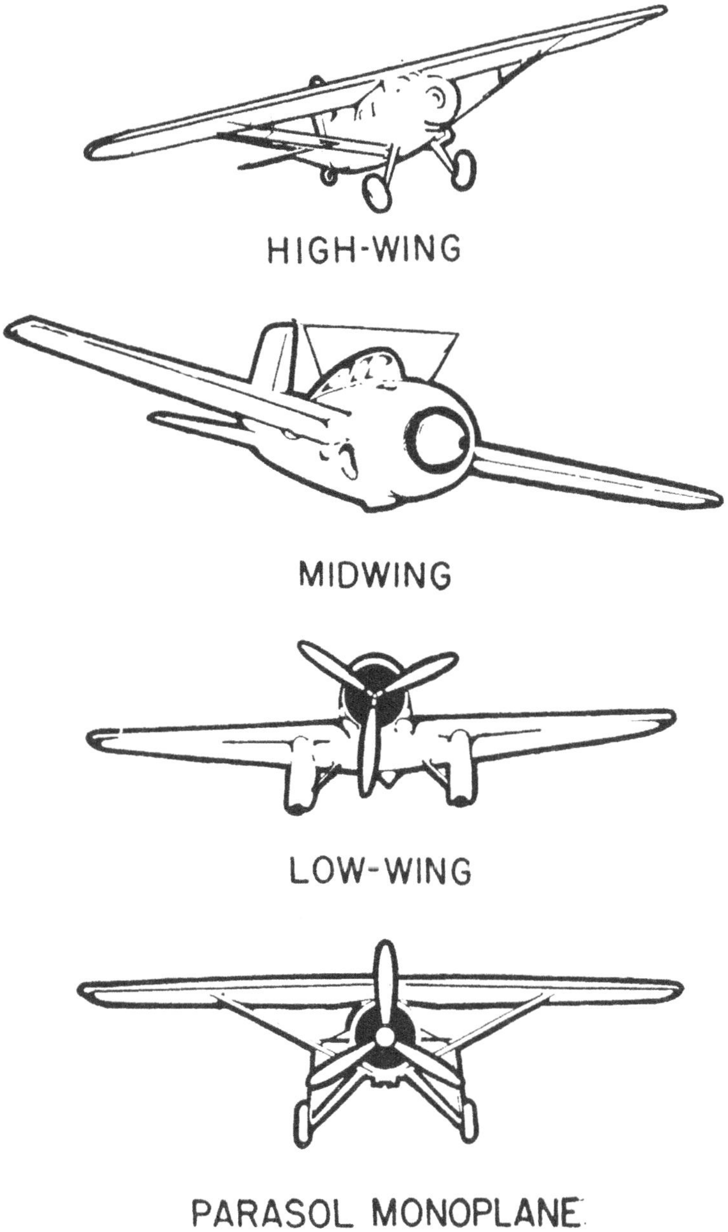

FIGURE 59.—Monoplane types.

also be a means of identification as a friendly or hostile airplane.

(3) *Characteristic features of individual airplanes.*—Knowledge of the features will permit you to determine whether an airplane is friendly or hostile, and when you are able to note them in sufficient detail, the particular model.

d. In general, you will identify airplanes by noting and combining indications under all three classes. In order that all possible indications may be quickly noted and evaluated, you must learn what characteristic indications are most likely to be detected under conditions existing at the moment you observe it. The study of FM 30–30, 30–31, 30–35, and 30–38, training films, and film strips on the identification of foreign and our own aircraft, and the habit of identifying every airplane you see or hear overhead will develop your observation.

■ 88. WARNING SIGNALS.—*a.* When you are posted as an antiaircraft lookout your major task is to warn your unit of an impending air attack or the approach of hostile observation airplanes in sufficient time to permit it to take the necessary protective measures. After your warning, the time available to your unit to disperse, take cover, and man antiaircraft weapons will be most limited. In most such instances you will have to decide instantaneously as to whether the airplane sighted is friendly or enemy and, if enemy, give the alarm. Unless you can positively identify the airplane as friendly you must give the alarm.

b. The prescribed warning of the approach of hostile aircraft, and of hostile mechanized vehicles, is as follows:

Use three long blasts of a whistle, vehicular horn, siren, or klaxon, repeated several times; or three equally spaced shots with rifle or pistol; or three short bursts of fire from machine gun or submachine gun. In daylight when you give the warning, if within sight of your unit, point in the direction of the airplane; at night, if within hearing, you must supplement your alarm signal by shouting the direction of the danger.

NOTE.—For additional information on the subject of this chapter see TF 1–259 and 1–294.

CHAPTER 5

ANTIMECHANIZED PROTECTION

SECTION I

ANTIMECHANIZED MEASURES

■ 89. GENERAL.—The speed and mobility of armored vehicles are such that you must always be prepared for their attack. Proper use of concealment and cover, training in the employment of the various antimechanized means available to you, and a thorough knowledge of the capabilities and limitations of the various types of armored vehicles will permit you not only to protect yourself and your unit against their attack but to take full advantage of your opportunities to destroy them and their crews.

■ 90. TYPES OF ARMORED VEHICLES.—In general, armored vehicles are classified as scout cars, armored cars, and light, medium, and heavy tanks. In addition to these, self-propelled assault, antitank, and antiaircraft artillery, and armored personnel carriers have some of the characteristics of armored vehicles.

■ 91. IDENTIFICATION.—It is most important that you study the pictures and silhouettes of both friendly and hostile armored vehicles. You must be able to distinguish at once between the enemy vehicles and our own, not only to avoid firing on friendly ones, but so that you can identify yourself and your unit to the occupants, if necessary, to prevent them from firing at you. You should pick out the most apparent characteristics of each type of hostile vehicle, the shape of the turret, how the weapons are mounted, the track features, etc. (fig. 60). Quick identification of the type of vehicle plus your knowledge of the characteristics of that type will enable you to take the most appropriate action against it.

■ 92. CHARACTERISTICS.—All of the above vehicles have the following advantages in varying degree. They possess considerable speed both on roads and across country. Their

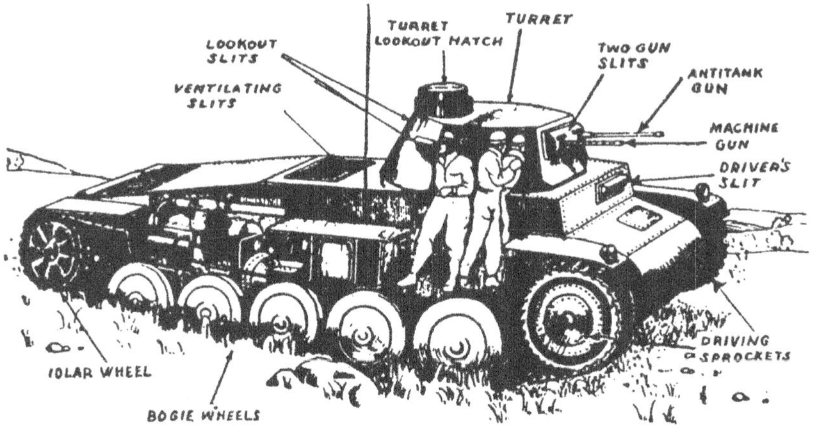

① Right side, front, and top view.

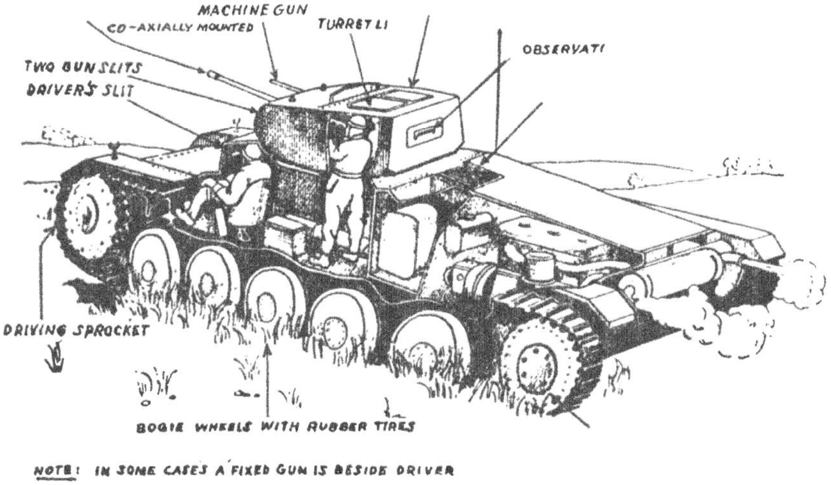

② Left rear view.

FIGURE 60.—Most apparent characteristics of tank and location of crew.

weapons are automatic or semiautomatic and they can deliver a heavy volume of fire. Their armor protects their occupants to some degree from hostile fire. These vehicles also have certain unfavorable characteristics which vary, as do their advantages, with the type of vehicle. These limitations are poor observation, noise, inaccuracy of fire while moving, sensitiveness to terrain, crowded interiors, and the bridging action of tracks.

■ 93. LIMITATIONS.—To secure maximum protection against armored vehicles and the best effect from your antimechanized weapons you must take full advantage of the limitations of these vehicles.

a. Observation.—The armor which protects the occupants of these vehicles limits observation. The more complete the armored protection, the more limited is the observation. When a tank, for example, is closed for combat, the vision of the occupants is limited to a few narrow slits in the armor and to the small field of view of its periscopes. The speed and jolting motion of armored vehicles also hamper observation. The occupants do not have time to study the terrain carefully.

b. Noise.—The noise in armored vehicles makes it difficult for the crews to hear sounds from outside the vehicle. The noise of the engine and track of a tank will so deafen the crew that they are frequently unable to hear the firing of hostile weapons, even antitank guns.

c. Inaccurate fire while vehicle is in motion.—It is very difficult to aim any of the weapons of an armored vehicle when the vehicle is in motion. Fire is quite inaccurate even when the vehicle is on a smooth road, and when it is moving across country accurate fire is usually impossible. Successful tactical employment of small groups of armored vehicles depends upon their ability to locate targets at sufficient distance to permit some vehicles to halt, preferably with chassis defilade, or in concealment, and fire at the target, while others maneuver to attack the target in flank or rear. Early disclosure of your position is what the armored vehicles hope for, and is sure to be disastrous to you.

d. Sensitiveness to terrain.—Many natural obstacles block the advance of armored vehicles (par. 95). In almost all terrain there are features which make difficult the cross-country movement of individual vehicles. Such natural objects as a pile of rocks, a deep hole, shell or bomb craters, a spring hole, a group of trees, or a stump-filled area will force a vehicle to change its course. The driver is constantly on the watch for these minor obstacles and, since his vision is limited, he is inclined to turn, when confronted with such an obstacle, and follow a neighboring vehicle that is able to move ahead. In small groups of armored vehicles this tendency to take a proved path results in frequent canalization and grouping.

e. Crowded interiors.—All armored vehicles have limited interior space and the occupants are crowded. The result is that if any of the fire of your weapons penetrates such a vehicle it is very apt to hit several or all of the occupants. The steel interior causes bullets and fragments of metal to ricochet or to disintegrate so that bullet splashes may spray the whole crew.

f. Bridging action.—The distribution of the weight of tanks and other track-laying vehicles permits them to pass over narrow trenches and holes without crushing in the earth. If you are below the surface of the ground in such cover, a tank will pass over you harmlessly.

■ 94. SECURITY.—Since the threat of armored attack is nearly always present, antitank lookouts will be provided by your unit to give warning of the approach of armored vehicles. You must be prepared to carry out this important duty. You will be posted where you can see an area of possible approach, and provided with sound or visual means of communication, or both, for warning your unit. Since armored vehicles are bulky and difficult to conceal, and because they make considerable noise and frequently raise unusual columns of dust, you will be able to give adequate warning to your unit.

■ 95. CONCEALMENT AND COVER.—Use of the methods of concealment discussed in chapters 1 and 2 will permit you to escape observation by the crew of an armored vehicle unless you disclose your position by movement or fire. Many natural obstacles block the passage of armored vehicles. Deep streams, canals, or other bodies of water, marshes or boggy ground, deep ditches or ravines, thick heavy woods, stump land, and ground littered with large boulders offer the best protection. A narrow trench or fox hole in firm soil, in which you can get below the surface of the ground, will also protect you from tanks and other track-laying vehicles (pars. 24 and 26).

■ 96. ANTIMECHANIZED WEAPONS.—Some or all of the following weapons will usually be available to you for use against armored vehicles. You must know how to use all of these, and know the best way to use them against the different types of armored vehicles. They are—

Armor-piercing ammunition.
Grenades, rifle and hand; high explosive, and smoke.
Antitank mines.
Explosives, TNT blocks, etc.
Obstacles.

■ 97. ARMOR-PIERCING AMMUNITION.—Armor-piercing ammunition is issued for all weapons from caliber .30 to 75-mm. The caliber .30 armor-piercing ammunition will penetrate the armor of a scout or armored car, the armored personnel carrier, and the shields of some self-propelled artillery. Although it will not penetrate tank armor it will go through the belly of many types of tanks and jam unprotected turret tracks and gun mounts. It will enter the vision slots, pistol ports, and sometimes rivet holes and ricochet or splash inside the vehicle. Caliber .50 machine-gun ammunition will penetrate the armor of many light tanks and cause the inside heads to fly off rivets on many other types. The rivet heads can inflict serious casualties. The heavier type of armor-piercing ammunition such as 37-mm or 75-mm will penetrate all but the heaviest armor on heavy tanks.

■ 98. GRENADES.—Hand and rifle grenades are capable of doing considerable damage to armored vehicles at close range. Such weapons can wipe out the crew of a vehicle when the explosion is confined to the interior. They can jam a turret, put a gun out of action, set fire to a tank, and sometimes break the track of a tank. Smoke grenades will blind the crews of armored vehicles and permit you to close with the vehicle with an explosive charge or grenade or to move to other cover.

■ 99. ANTITANK MINES.—These mines are designed to break the tracks of tanks. They will wreck a wheeled vehicle. To be effective they must be concealed and their location covered with fire to prevent their removal. They are particularly effective when the possible approaches to a position are limited (par. 95).

■ 100. Explosives.—Improvised mines may be made from properly fuzed TNT blocks. Such explosives are excellent for destroying a tank which has been stopped either by damage or by smoke.

■ 101. Obstacles.—The construction of antimechanized obstacles to mechanized attack is covered in FM 5–30. You will normally construct such obstacles under the supervision of an officer or noncommissioned officer. You must remember that no obstacle is effective unless covered with fire to prevent its removal.

■ 102. Combat With Armored Vehicles.—*a*. Ingenuity coupled with aggressiveness will permit the individual or the small group to give an excellent account of themselves in combat against single or small groups of armored vehicles.

b. Study the terrain about you from the point of view of the crew of an armored vehicle. Plan the action which you will take in the event it uses any possible approach. Make full use of the presence of large boulders, stumps, and trees. Your concealment should include all measures by which you can remain concealed while firing at the vehicle. In dry ground the dust raised by the blast of your weapon will be more apparent than the flash. Such ground should be wet down if possible, or covered with wet sacks, mats, or small branches, leaves, or grass. Select nearby alternate positions to which you may move if the enemy discovers your first position. Take full advantage of the presence of large boulders, stumps, trees, ditches, and other tank obstacles. If your position is discovered, some vehicles will fire on it while others try to approach by covered routes from the flank and rear. Study such approaches and plan your action to counter such a move. Locate or dig cover so that a tank cannot run over you.

c. Hold your fire until it will be effective. The success of your efforts will depend upon the surprise with which they are made. All the members of your squad must exercise self-control to avoid premature disclosure of their positions. Many armored vehicles may pass close to your position but beyond effective range of the offensive weapons at your disposal. Let them go; others will get them. Wait for your opportunity to take action against the vehicles or troops which follow. If a tank is disabled or stopped near you, don't open fire just to hit the tank. Remember that the crew still has protection and can use their weapons. Wait until the crew emerges and pick them off with rifle fire, or approach the tank by a concealed route and wait for an opportunity to use a grenade or explosive charge.

■ 103. Ambush.—*a*. There will frequently be opportunities to ambush armored vehicles. In most terrain, natural obstacles, not apparent on maps or air photographs, form short defiles and canalize small groups of tanks (par. 92). Such natural obstacles usually afford excellent concealment and cover for your ambushing group. You should scatter your group throughout the length of the defile. Antitank mines, mounted on a plank or other means so as to be easily pulled over the ground, should be hidden and drawn under the tanks as they pass. Some mines may be laid to block the exit from the defile and others arranged to be drawn under tanks after the leading tank is stopped. When only small arms and grenades are available, self-control will have to be used and your future action will depend upon what the uninjured tanks do. Watch for opportunities for well-aimed shots at the periscopes, the vision slits, or at the turret track. If you are close enough to make your fire effective, you will be close enough to see what the crew members are doing. If the turret guns are pointing in your direction, remain concealed. If you have fired and the turret turns toward you, shift quickly to your alternate position and let someone fire on the tank from another direction. If smoke grenades or candles are available use them to blind the crews and you can approach close to the tank. Remember the smoke will be drawn into the tank by the ventilating system, so watch for the turret to open and use your rifle, pistol, or hand grenade. If several tanks are blinded by the smoke and try to move they may put themselves out of action by collision and the crews may be destroyed when they attempt to leave the crippled tanks.

b. Because of limited observation, tanks close their turrets only when forced to do so by the threat of your fire. In woods or heavy brush the tank commander will frequently observe with the turret open. Such terrain affords you good concealment and you will frequently be able to get so close that you can shoot the tank commander before he can close the turret. A comrade, on the opposite side of the tank's path, should be prepared to drop a hand grenade in the open turret.

c. Scout and armored cars and other wheeled armored vehicles, which have less cross-country mobility than track-laying vehicles, use roads or trails whenever possible. Road cuts, wooded areas, and trails through swamps are excellent places to ambush such vehicles. Any armor-piercing ammunition will be effective against these vehicles. All open-topped vehicles are ideal targets for grenades. The fragments which do not inflict casualties on explosion will usually do so on ricochet. Your plan should provide for some action to attract the attention of the car's crew when the car is opposite your ambush. A burst of fire from a point down the road, a tree dropped in the road, even the waving of a flag will usually be sufficient to attract the attention of the occupants of the vehicle and cause the vehicle to slow down or stop. At such a moment an attack from the rear by fire and grenades may be a complete surprise. Provision must be made for security from the action of following vehicles.

Motorcycles frequently precede armored vehicles to discover road blocks and traps, and to draw fire. Frequently it will be best to let the motorcyclists pass your ambush, making provision for another group to get them or making a plan to get them when they return. If that is not practicable try to kill or capture them without noise and thus avoid warning the following vehicles.

Motorcycles cannot run slowly for very long without burning out their motors. They usually advance by short, fast rushes, along roads, from one crest or road bend to the next. Take advantage of this habit in planning your ambush. Set your trap between such points. A rope or piece of telephone wire stretched diagonally across a road may be used to throw a motorcyclist. It should be fastened so that there is enough slack to conceal it where it crosses the road. On hard-surface roads it may be concealed by making it appear merely part of litter, including other pieces of rope or wire, scattered on the road in that vicinity. A jerk to pull it to the desired height just before the motorcyclist passes over it will be sufficient to throw him into the ditch. One man should be concealed close to the road prepared to quiet the rider if the fall doesn't accomplish it. On narrow roads a small tree pivoted on one side may be pulled across quickly to accomplish the same result. Such a tree should be concealed by being one of several such trees lying along the road. Oil on a high-crowned road or at a curve will often cause motorcycles and wheeled vehicles to skid into the ditch or overturn.

d. Ambushing and stalking armored vehicles offer full play to individual initiative. Ingenuity will find many ways of overcoming the initial advantages of the crew of an armored vehicle. Hunting armored vehicles is the big-game hunting of modern war. Intelligent use of concealment and cover, a thorough knowledge of the quarry's weaknesses and habits, and expert and aggressive use of your weapons will give you "good hunting."

SECTION II

ANTIMECHANIZED LOOKOUTS

■ 104. GENERAL.—No matter where your unit is in the theater of operations it must be prepared to meet a hostile mechanized attack. Adequate warning of the approach of armored vehicles is vital. To prevent surprise, antitank lookouts must be posted in positions to give warning to the unit. You must be prepared to act in this important capacity.

■ 105. LOCATIONS.—Antimechanized lookouts must be able to see all possible avenues of approach for such vehicles. When your unit is a considerable distance from the enemy the most probable avenues of approach will be roads. In the combat zone, mechanized attacks like air attacks must be expected from any direction hence all-around security is necessary. When on duty as an antitank lookout you must be able to see the approaches in the area for which you are responsible and not be seen by hostile forces. Never take a position in a road or stand in an open field. Use the concealment measures discussed in chapters 2 and 3. In addition to concealment you should provide cover for protection in the event the hostile mechanized forces move over your position (par. 95)

■ 106. DUTIES OF LOOKOUT.—Your mission as an antimechanized lookout is to warn your unit of the approach of hostile armored vehicles so that it can take the measures which have been planned to meet such an attack. To do this you must—

a. Be always on the alert to hear and see all indications of the approach of armored vehicles.

b. Be able to identify all vehicles which are moving in the area for which you are responsible.

c. Warn your unit of location, direction of movement, and, if possible, the strength of the hostile force.

■ 107. INDICATIONS.—Armored vehicles are noisy and the sounds they make are quite distinctive. You will frequently hear them before you see them, particularly in close terrain. These vehicles raise very heavy clouds of dust in dry country. Since there will frequently be many friendly vehicles in your vicinity, sound and dust clouds alone will not be sufficient cause for giving the alarm unless your orders direct it or when your area of observation is toward ground occupied by the enemy. These indications should never be neglected, however, and you should be particularly watchful in the direction from which they come.

■ 108. IDENTIFICATION.—The identification of armored vehicles demands study and practice. Field Manuals, training films, and film strips are available for this purpose and you will be trained in this identification. You should know thoroughly all the types of vehicles which are used in our Army and in those of our allies. Consider all others as hostile vehicles. Since tanks are frequently partly hidden by their own dust clouds you should practice until you are able to identify a tank by a partial view.

■ 109. WARNING.—Depending on the situation you will have visual, audible, or other means of giving a warning. The warning signals given in paragraph 88 are those used for both hostile aircraft and hostile mechanized vehicles. In addition to these you may be furnished with pyrotechnic signals, telephone, or radio. Be sure that you understand fully the methods which you are to use as your warning is vital not only to your unit but perhaps to the entire command of which it is a part.

NOTE.—For additional information on the subject of this chapter see FM 5-30, 30-40, and 30-42; TF 5-145, 5-146, 5-147, 5-148, and 5-149; and FS 5-6.

CHAPTER 6

PROTECTION AGAINST CHEMICALS

■ 110. GENERAL.—Chemical agents are scientific weapons which must be countered by scientific methods. Their effectiveness is primarily dependent on surprise. The gas mask and the protective clothing issued to you provide a scientific protective means. The training against chemicals which you receive, plus the enforcement of individual and unit gas discipline, will prevent surprise and effectively counter hostile use of chemical agents.

■ 111. FEAR OF CHEMICALS BASED ON IGNORANCE.—In the first World War more than one-quarter of the total casualties in the United States Army was due to gas and of all these casualties only 2 percent died. This occurred in the days when protective clothing was poor or nonexistent and the troops, at least in the early stages, had little or no gas training or gas discipline. Equipped with the excellent mask and impregnated clothing available today you must always be on the alert against chemical agents, know how to recognize them and how to protect yourself against them, and thus minimize their effectiveness as casualty-producing weapons.

■ 112. TRAINING AND DISCIPLINE.—Chemical weapons are effective only against the unprotected and the unprepared, the lazy and the careless. They will not pardon any carelessness, any ignorance, or any deficiency. The effects of chemicals on the victims may be terrible, but such casualties among troops will be almost entirely due to carelessness. Chemical agents never have won a battle against trained and well-equipped troops. Serious casualties produced by gas are well below the average risks of battle, provided that the troops possess good gas discipline. By this is meant that the individuals are on the alert, can recognize the various chemical agents and counter them by the prompt wearing of the gas mask, undertake appropriate decontamination work at the right time, and, lastly, render such first aid as may be required.

■ 113. Gas Mask.—You have been trained in the proper adjustment and drilled in the use of the gas mask. You have received instruction in the identification of the commonly used chemical agents. In time of war you will be issued a gas mask fitted to your particular needs, and protective clothing. The Army gas mask, properly used, gives 100 percent protection for the eyes and lungs, and the protective clothing the same against the vapors of mustard gas. Frequent opportunities to perform your normal duties while wearing the mask will accustom you to it. The introduction of chemical warfare features into your normal tactical exercises will develop gas discipline in your unit.

■ 114. Methods of Attack.—A chemical attack may be made in one of the following ways:

a. From candles and cylinders.—This method may be apparent during daylight by the visible gas cloud, or in some instances at any time by the hissing sound of the gas escaping from cylinders.

b. From gas projectors, artillery and mortar shells, and chemical grenades.—Projector attacks make a big explosion, a brilliant flash, and a large cloud of smoke and dust. Artillery and mortar shells and grenades filled with chemical agents sound almost like duds when they explode. Usually a thin haze or mist surrounds the burst for a few moments.

c. From airplanes and tanks.—The airplane bomb filled with gas also sounds like a dud when it explodes. If the gas is sprayed from the airplane or tank it can usually be seen.

d. From bulk containers and chemical land mines.—These are placed in position and exploded by electricity or by contact fuze.

■ 115. Gas Officers and Noncommissioned Officers.—Each regiment and battalion has at all times a gas officer and a gas noncommissioned officer, and each company (battery, troop) has two gas noncommissioned officers. The primary duty of these individuals is training in protection against chemical attack. Thus expertly trained personnel is available at all times to assure the maintenance of gas training, of protective means, and of gas discipline.

■ 116. GAS SENTRIES.—Wherever there is a threat of gas attack, gas sentries are posted for the protection of all members of the command. A gas sentry always guards troops which are working or sleeping. You must be prepared to serve efficiently in this important capacity. The assurance that they will be warned of a chemical attack will contribute greatly to the morale of your comrades.

■ 117. DUTIES OF GAS SENTRIES.—*a.* Your duties as a gas sentry are to—

(1) Enforce all specific orders of your post for defense against chemical attack.

(2) Locate the position of all sleeping men in your area.

(3) Detect the presence of chemical agents by odor, color, and appearance.

(4) Give the alarm whenever gas is detected.

(5) Detect sounds indicating the preparation for and actual enemy projection of chemical agents.

(6) Protect supplies in your area.

b. In the event of a gas attack, you must adjust your mask, sound the alarm, and immediately proceed to wake up every man in your area. You will not pass on alarms arising in other areas but will give the alarm only when you yourself detect the presence of gas. This precaution is necessary in order to prevent the spreading of false alarms. Standing orders of your unit will prescribe your duties as a gas sentry in cases of general gas alarms, and it is the duty of your unit gas officer to insure that you understand these instructions.

c. Provision must be made for posting an adequate number of gas sentries over sleeping troops and working parties in the theater of operations to insure that every sleeping man shall be wakened and every man engaged in work warned in time. As a gas sentry, always on the alert to detect the presence of gas, you provide security to the other men of the command and permit them to gain much-needed sleep and rest. This is your most important function. You should be equipped with a gas alarm, should have a definite area to alarm, and should know when to give the alarm.

d. You may be detailed as a gas sentry at a gasproof shelter. Take your post on the up-wind side. In the event of a gas attack, you perform the following duties:

(1) Adjust your mask and sound the alarm to warn occupants of the shelter to close inner door or lower the blanket over the entry.

(2) Close outer door or lower blanket.

(3) Look after proper entrance and exit of personnel.

(4) Assure the airtight adjustment of outer door or blanket.

(5) Require all the prescribed degassing measures of personnel before entering.

(6) See that number of entrants does not exceed prescribed capacity of a shelter.

(7) Direct starting and operation of collective protector at a ventilated shelter.

(8) Open door or raise blanket for litter bearers at an aid station.

(9) Test for gas and advise occupants of a shelter when it may be opened after a gas attack.

(10) After a gas attack, direct that air lock and shelter, if necessary, be cleared of any gas that may have entered.

(11) Resume your post promptly.

■ 118. Special Gas Sentries.—These sentries are usually posted to guard dangerous gassed areas or supplies at distributing points. Their duties do not differ from the usual gas sentry except that they are given special instructions regarding special duties. If you are detailed as a special gas sentry on duty where supplies are stored you should have protective clothing and a mask. Such installations will be chiefly the targets of enemy attack employing spraying devices. Your primary duty is to protect the supplies. Since such an air attack develops very fast, you have only a few seconds to pull the protective covers into position. You should, therefore, always be ready to act at an instant's notice. If the enemy air attack includes incendiary bombs, you will give the prescribed fire alarm.

■ 119. Rules to be Remembered.—You should remember the following rules which will assure your protection against chemical attack:

a. Do not carry anything in your gas mask carrier but the mask and tube of "antidim."

b. Do not neglect the gas mask or allow it to receive rough handling. Never use your mask or carrier as a seat or pillow.

c. Do not throw away your gas mask. You may need it later on and it will save your life in a gas attack.

d. Do not give a false gas alarm.

e. Do not breathe after the gas alarm is given until you are sure that your mask is well adjusted to your face and that the facepiece has been cleared of gas by blowing vigorously into the facepiece while holding the outlet valve.

f. Do not remove your gas mask until permission to remove it is given by an officer or a gas noncommissioned officer. (Violation of this rule was a major cause of gas casualties during the first World War.)

g. Do not enter an unprotected dugout immediately after a chemical attack.

h. Do not talk or move about unnecessarily during a gas attack.

i. Do not become panicky; keep calm and remember your protective equipment is effective if properly used.

j. Do not fail to realize that the enemy uses many different kinds of gases, sometimes alone, at other times mixed with other chemical agents, smoke, or high explosives.

k. Do not forget that clothing contaminated with blistering (vesicant) gas should be removed as soon as possible, and must not be used again until decontaminated.

l. Do not remove another man's clothing or handle equipment that is contaminated with liquid blistering gas unless you are equipped with protective gloves and, if possible, with complete protective clothing.

m. Do not forget that mustard gas or other vesicants may remain in an area for days.

n. Do not enter an area contaminated with blistering gas unless equipped with protective clothing and gas mask.

o. Do not remain for any length of time in an area contaminated with blistering gas, even if equipped with protective clothing and gas mask, unless required by the tactical situation.

p. Do not fail to post a gas sentry over sleeping men.

q. Do not forget that when the wind is blowing from the enemy between 3 to 12 miles per hour, a cloud chemical attack from the enemy may be expected.

r. Do not forget that during a calm, in foggy or cloudy weather, and at night, ideal conditions exist for a chemical attack. Be on the alert.

442577°—42——9

s. Keep on the up-wind side of areas where you think gas may be present.

t. Do not bathe in streams or lakes contaminated with chemical agents.

u. Do not allow men to drink water or eat food contaminated with chemical agents, or to use matériel which has not been decontaminated after being exposed to chemical agents.

v. Do not forget that all gas cases require, first, rest; second, warmth; third, fresh air.

w. Do not permit men who are casualties from inhaling gas, to walk, talk, or move about.

x. Do not bandage the eyes of a gas casualty; blindness may result.

NOTE.—For additional information on the subject of this chapter see FM 21–40; TF 3–216, 3–217, 3–218, and 3–219; FS 3–1, 3–3, and 3–7.

CHAPTER 7

PROTECTION OF MILITARY INFORMATION

■ 120. General.—Any information of a military nature which the enemy can obtain may help him in defeating our actions against him. Facts which may in themselves seem quite trivial may furnish the details necessary to give him our plan of action. You must be on your guard at all times against careless words which might be the cause of injury to yourself and your unit.

■ 121. Secrecy Discipline.—*a*. The enemy may have a highly organized spy system, against which all military personnel must constantly be on guard. Even though we guard against this system, it will never be broken up completely. You must realize that thoughtless or talkative persons, those who seek to impress others with their importance by showing the amount of information they possess, and those who feel they must tell family or friends all they know, may become a menace to the country and to the lives of their comrades (figs. 61 and 62)

b. You should never discuss military instructions, plans, operations, movements, the composition or location of troops, casualties, morale, military equipment or supply, results of hostile action such as sabotage, air attack or gunfire, or any other military subject in the presence or hearing of any stranger, either military or civilian. Such stranger may be a friend of our enemies. It is unwise to reveal even the most insignificant matters to strangers, for, by piecing this information together with that obtained elsewhere, they may deduce matters of the greatest importance (fig. 63.) If you hear others discussing military information in public places you should caution the offenders, and, if they fail to heed the warning, get their names and report them to your commanding officer or your unit intelligence officer. You should report to your commanding officer any person whose actions or conversations are such as to arouse your suspicions.

① "I tell you the whole --th Division is going to be billeted here."

② It *was*. (The consequence of careless talk.)

FIGURE 61.

① "Jack writes that his regiment is sailing next Monday."

② Next Tuesday. (Don't discuss troop movements!)

FIGURE 62.

① "Yeah, I was lucky. Why we went

② "Mister, we don't discuss military operations!"

FIGURE 63.

■ 122. SECRECY IN SIGNAL COMMUNICATION.—*a*. In time of war, secrecy in the transmission of messages is of the utmost importance. You must observe strictly all orders which are given you concerning secrecy of messages. You should never use a public telephone to discuss secret or confidential matters since one of the telephone operators may be an enemy agent (fig. 64). In a *stabilized* situation it is unwise to use the field telephone in forward areas for important conversations, unless time is vital. By induction or wire tapping, the enemy in nearby positions may hear all that is said.

FIGURE 64.—"The ----th Brigade will move to Ord by truck tonight."

b. The following are some of the more common ways in which our messages may be obtained by the enemy:

(1) Interception of radio communication.

(2) Interception of wire communication by tapping telephone and telegraph circuits directly or inductively.

(3) Capture of messengers carrying messages or codes, ciphers, or cipher keys.

(4) Activities of spies at message centers or other offices where messages are handled.

(5) Interception of visual signals by observing lamp, flag, pyrotechnic, and panel signals.

① "So that's where they're going. Fritz will like to know that."

② "The bums never leave nothin'."

FIGURE 65.

(6) Capture of codes, ciphers, and keys at a headquarters.
(7) Shooting of pigeons and capture of their messages.
(8) Treason.

■ 123. TROOP MOVEMENT PRECAUTIONS.—*a*. Before leaving a camp, concentration area, rest area, bivouac, or any other assigned area in the theater of operations be sure that you have not left behind any documents, signs, letters, or equipment. Such things might fall into the hands of the enemy or his agents and give him important information (fig. 65).

b. In order that there may be a check upon the movements of individuals in the theater of operations, restrictions will be placed upon the individual movements of civilian and military personnel. This is necessary to secure control of enemy spies. When absent from duty in your organization you must carry a furlough or pass, or a copy of the order directing your individual travel.

c. During war, when your unit is moving by motor, rail, or water, it is forbidden to mark trucks, cars, or baggage to indicate the organization, the destination, date of departure, or name of the ship in an oversea movement (fig. 66). You must be particularly careful that nothing you say or write during the movement will reveal your organization or its destination.

■ 124. CONTROL OF PERSONAL CORRESPONDENCE.—*a*. During war, letters, even those written home, are apt to fall into enemy hands. In order that no letter originating in the theater of operations will give facts of value to the enemy, all postcards, letters, parcels, cablegrams, telegrams, or radiograms of all military personnel originating in the theater of operations must be censored. No one likes this censorship, but past experience has shown that it is necessary for your own protection and the protection of your fellow soldiers. Your company commander will give instructions about censorship. In the theater of operations all private communications will be transmitted only by means of authorized systems of communication. For military personnel, this means the Army post office. You must never under any circumstances carry, or ask others to carry for you, letters outside the combat zone. These restrictions are necessary so that information will not fall into enemy hands (fig. 67).

① "Hey, Mame, we're sailing Saturday!"

② Conceal identifying marks and keep your mouth shut!

FIGURE 66.

① WRONG—"Say, Bud, mail this for me, will you? The censor always holds them up."

② RIGHT—"Sergeant, will you have this censored for me, please."

FIGURE 67.

b. Correspondence with enemy personnel, even with prisoners of war in our hands, is prohibited. If you wish to correspond with United States prisoners of war you should consult your commanding officer for the proper address.

■ 125. DIARIES.—As private diaries may contain information of value to the enemy, it is forbidden to send them through the mail. If you write a diary and keep it in your pocket or kit, it may be captured by the enemy and thus it is a source of danger. To minimize such risk, your organization will collect such diaries periodically, enclose them in sealed covers marked with the name, grade, and organization of the writer, and send them to the rear to be stored. You will be instructed how to make proper application to recover such a diary. Don't carry any part of a diary into battle.

■ 126. PHOTOGRAPHY.—The use of cameras must be restricted in time of war. It will generally be forbidden to take a privately owned camera into the combat zone.

■ 127. SAFEGUARDING MILITARY INFORMATION.—By safeguarding military information is meant those measures which are taken to protect vital military information from falling into the hands of possible enemies of our country. Military information is safeguarded both when we are at peace and at war. Wherever you may be, there may be enemy agents and residents of the area who will divulge to the enemy any information of a military nature that comes to their attention. Accordingly, it is imperative that you exercise great discretion in your conversation and care in handling documents containing military information. Such documents are safeguarded by classifying them as secret, confidential, or restricted, depending upon the nature of their contents. At times you may be issued certain classified documents. You must safeguard them and allow no unauthorized person to see them. If in doubt as to the authority of a person who wishes to see a classified document, don't show it without permission from your commanding officer. If you find a classified document you should deliver it to your organization commander. In addition to documents, much matériel is safeguarded. You should be on your guard against unauthorized persons endeavoring to examine or photograph equipment or weapons, information of which is not generally distributed to the public.

■ 128. ACTION PRIOR TO GOING INTO COMBAT.—Prior to going into combat, all distinguishing marks and insignia on vehicles, equipment, or persons will be removed or obliterated under an officer's supervision. You must be careful to remove any divisional, regimental, or company insignia from your clothes and equipment. Search your pockets for letters, memoranda, orders, souvenirs, or keepsakes which might disclose your organization. All such articles, together with diaries, must be turned in for safekeeping when you are going into the combat zone.

■ 129. ACTION IN CASE OF CAPTURE.—*a*. One of the best sources of information is that obtained from prisoners of war. One uninstructed soldier, if captured, may betray important information to the enemy by his answers to seemingly innocent questions.

b. If you are captured by the enemy, he will make every effort to question you. The enemy will not only resort to direct questioning, but may place agents in our own uniforms in prisoners-of-war cages, and conceal microphones in buildings occupied by prisoners for the purpose of recording conversations. In the event of capture, observe the following rules:

(1) Give correctly your name, grade, and serial number, and maintain absolute silence when asked any other questions (fig. 68). This is all you are required to give by International Law and you cannot be punished for refusing to give further information.

(2) Under no circumstances give untrue or misleading answers. The enemy interrogator will probably know enough to be aware that you are not telling the truth. You can be punished for lying but not for refusing to answer unauthorized questions.

(3) While you are a prisoner of war do not discuss at any time or place anything concerning your country's armed forces.

(4) Do not give the name or number of your organization. Should capture become imminent, destroy any messages you may be carrying, and search your pockets again to destroy

① "Note that! He's from the ______ Infantry of the ______ Division located at Bouillon."

② "James Doyle, Corporal, serial number 98,765,432, same as it says here."

FIGURE 68.

anything which may be of value to the enemy's intelligence service.

■ 130. RIGHTS OF PRISONERS OF WAR.—*a*. You should know the rights of prisoners of war not only to know what you may expect if you are captured but the rights of prisoners which you may capture. Prisoners of war are in the power of the enemy but not of the individuals or bodies of troops who capture them. Prisoners must at all time be treated humanely and be protected, particularly against acts of violence, insults, and public curiosity. Measures of reprisal against prisoners are prohibited. Prisoners of war have the right to have their persons respected.

b. All effects and objects of personal use—except arms, horses, military equipment, and military papers—as well as metal helmets and gas masks will remain in the possession of prisoners of war. Sums of money, watches, or jewelry in the possession of prisoners may not be taken away from them except by order of an officer and after the amount is determined. A receipt shall be given. Sums thus taken away shall be credited to the account of each prisoner. Identification tags and cards, insignia of grade, decorations, and objects of value may not be taken from prisoners.

c. Nations at war must notify each other as soon as possible, through an intermediary, of every capture of prisoners. They are likewise bound to inform each other of the official addresses to which the correspondence from the families to prisoners of war may be sent. As soon as possible, every prisoner must be enabled to correspond with his family himself.

■ 131. PROPAGANDA, HOW TO RECOGNIZE IT AND HOW TO COUNTERACT IT.—*a*. You should be able to recognize propaganda disseminated by agents of an enemy government. This propaganda may be spread by newspapers, articles in magazines, pamphlets, news pictures, motion pictures, radio broadcasts—including radio broadcasts at the front—and rumors by word of mouth through undercover sources. This propaganda will seldom be obvious out-and-out favorable statements about the hostile government or its military forces, but will usually be a false interpretation of known facts in an attempt to create in your mind doubt and distrust of your own Army and your own Government. Propaganda, to be effective, must contain many facts which the individual to be influenced knows to be true. It is difficult to recognize propaganda for what it is, and you should be constantly on your guard since what you read, see, or hear from unknown sources seldom tells the whole truth of the subject matter discussed. If it is found that the news obtained originated in a hostile country or even in a neutral country, if the author is unknown or if he is known to be unfavorable to our cause, if the news is a prediction of future events adverse to our cause, or if the facts stated do not agree with known truth about the matter, be suspicious of your information and label it as propaganda which is not to be believed.

b. The sole object of enemy propaganda is to lower the morale of our Army and of your people at home in order to insure the defeat of our armed forces, of which you are a part. You must resist any such attempt and by your own reason and patriotism maintain your confidence in the power of our people to uphold our Government and American ideals against any foreign power in the world.

NOTE.—For additional information on the subject of this chapter, see FM 30–25 and TF 11–204 and 11–225.

CHAPTER 8

PROTECTION AGAINST CARELESSNESS

■ 132. General.—Careless or thoughtless acts lead to frequent casualties in the combat zone. Such casualties have just as serious an effect on our fighting strength as battle casualties. You must always be as alert against accidents as against enemy action.

■ 133. Booby Traps.—This is the name given to devices used by an enemy to wound or kill careless or thoughtless opponents. As the name indicates, only a "booby" will be injured by such a trap. You must always be watchful for them in areas which have been in the possession of the enemy. They may be connected to doors, windows, furniture, and souvenirs, such as helmets or parts of uniforms or rifles, and they may be attached to weapons or vehicles which have had to be abandoned.

a. Description.—The elementary booby trap usually consists of a small charge of explosive, either in slabs or in a cartridge, with an igniter worked by either the push or pull methods or both (fig. 71). A wire or cord may be attached to the igniter and to another object or the trap may be placed beneath a loose board.

b. Location.—The following are examples of places in which booby traps may be used:

(1) *Doors.*—Traps are placed inside houses, over doors, in door frames, or against the wall behind the door when wide open. Doors in such cases are rarely locked and are sometimes left ajar (fig. 69①).

(2) *Windows.*—Traps are placed on inside sills of ground floor windows or connected to closed or half-open shutters with a wire in tension (fig. 69②).

(3) *Barn doors.*—Traps, usually of heavier charge than referred to in (1) and (2) above, may be placed on the ground or between battens at the height of a man (fig. 69③). They may be placed under a loose board in the floor.

(4) *Vehicles.*—Traps may be placed inside the cab of a vehicle, attached to the door or a brake handle, or under the seat. They may be found under the closed or half-open hood of the vehicle.

(5) *Souvenirs.*—A helmet, pair of boots, blouse, rifle, or pistol, or other piece of equipment, particularly when left in a building or other confined space, may be connected with a booby trap. This is a particularly useful type against an opponent with the habit of collecting souvenirs (fig. 69④)

(6) *Miscellaneous.*—Traps may be used in cellar entrances, inside doors, cupboards, chests of drawers, and farmyard vehicles.

■ 134. ANTIPERSONNEL MINES.—Such mines work with shrapnel effect, that is, use a casing containing hundreds of bullets or other missiles which are blown through the air by the explosion of the mine. Like booby traps they may work on the push or pull method (fig. 70) They are particularly useful in woods where they are more easily camouflaged. Mines with pull igniters are usually worked by a simple trip wire but may be fixed to loose branches and other minor hindrances to advance which an unsuspecting person is liable to move. A mine with a push igniter is generally used in combination with an obstacle, being placed in the middle of the obstacle itself or in the approaches to it. Antipersonnel mines may frequently be employed to protect barbed wire, antitank mine fields, and other obstacles from clearing operations.

■ 135 ANTITANK MINES.—Antitank mines may be found in small fields blocking roads or approaches to important weapons or positions or in larger fields used as a part of a continuous obstacle against tank movement. Many antitank mines operate only under pressures of 300 pounds or more but others may operate under the weight of a man or may be fitted with special means of detonation when tampered with. These mines should be neutralized and disarmed by specially trained individuals. If you encounter an area where you suspect their presence avoid it and report it to your commander.

■ 136. PRECAUTIONARY MEASURES.—*a.* When you suspect the presence of booby traps or of mines you must take great care

① Connected to door. ② Connected to window.

③ Be suspicious of sliding doors. ④ Souvenir hunter, beware!

FIGURE 69.—Booby traps.

① Pressure type.

② Pull (trip) type.

FIGURE 70.—Antipersonnel mines.

before moving. You should examine outdoor areas for trip and tension wires or disturbed surfaces of the ground. In buildings nothing should be moved until it has been thoroughly examined. If you must enter a building, try to get a view through a window of the inside of the door before you enter. If you can't do that open the door with a long pole from a sheltered position. Even if you thus detonate a mine be careful that there isn't a second one which will explode when you close the door. Sometimes a double booby trap has been used to take advantage of such carelessness.

b. When any wire or cord is found you should follow it up, *without touching it,* to make certain that it is not connected to a trap. Any igniter which you find must be neutralized before any wires are cut or any unnecessary movement is made. If you cut a cord or wire, it should be done with scissors or wire cutters, not a knife, so that no pull will be produced while cutting. If the wire or cord is stretched tight, it must not be cut as that will detonate the mine. If you cannot easily neutralize a mine or trap, mark it to protect others until it can be dealt with by specially trained men.

■ 137. NEUTRALIZING A TRAP.—The first and most delicate operation is to find out where the trap is and how it is fired. Having discovered the charge, all igniters must be neutralized. Push and pull igniters have a safety device consisting of a metal key which passes through a hole in the stem. The last step in placing the mine is the removal of this key. When you have found the igniters, you must neutralize them by pushing a split pin or nail in this hole (fig. 71). Until this has been done you should exercise great care to avoid compressing the igniter in any way or pulling any wires or cords in the vicinity. Once split pins are in position you should unscrew the igniter from the charge. Do this with care as a detonator may be in the igniter tube and, if so, it is liable to fall out. If you cannot unscrew the igniter by hand, leave it for a trained man. *Do not attempt to disarm a mine unless your mission demands it.*

■ 138. DUDS.—You will frequently find unexploded shells and bombs in the area in which you are working. Never examine them or pick them up unless you have been specially trained to do so and your particular mission is to destroy them. If

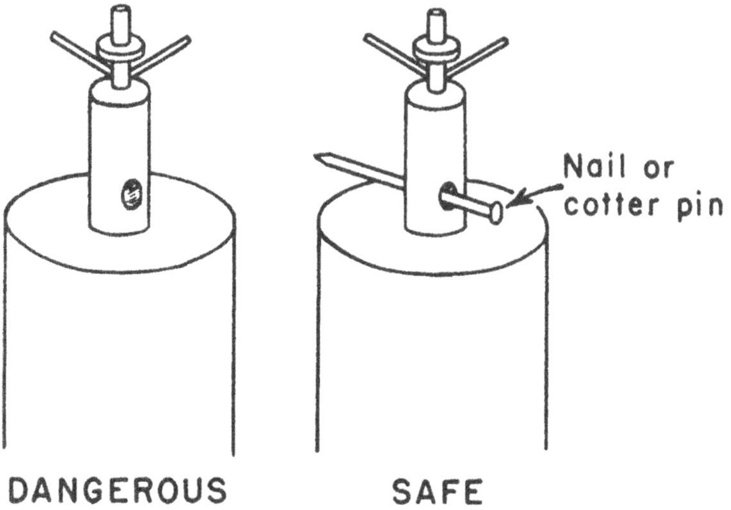

① Push igniters. ② Pull igniters.

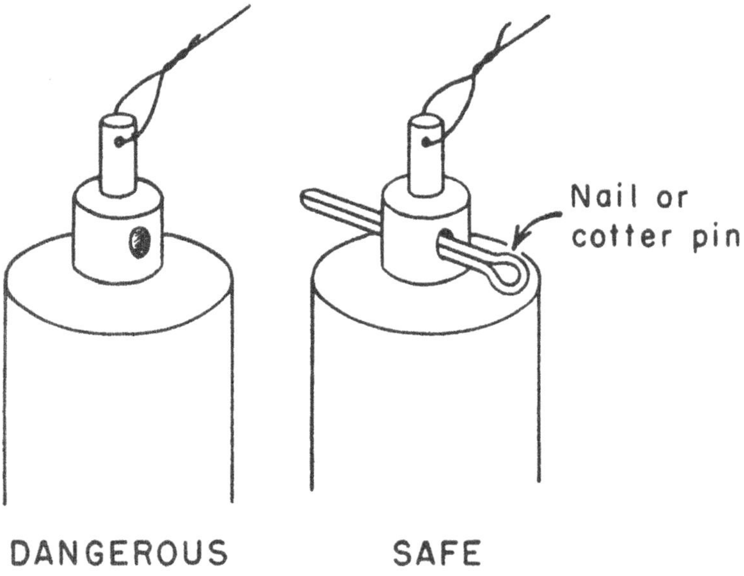

FIGURE 71.—Neutralizing a trap.

you discover such shells or bombs you should warn others of their presence and notify your commander of their location.

■ 139. MOTOR VEHICLES.—Carelessness in the vicinity of motor vehicles is a cause of many avoidable accidents. Casualties produced in this way are just as helpful to the enemy as if he had inflicted them himself. Do not expect a vehicle driver to be able to avoid an accident. He has a big job to do that demands all his attention and your cooperation. Never walk on a road if you can avoid it. When it is necessary, walk on the left, facing traffic. When you fall out to rest by the road, move off the shoulder beyond the ditch. If you are lying in the grass at the edge of the road it is your fault, not the driver's, if he hits you when suddenly forced to pull off the road. *Never go to sleep under a motor vehicle,* even if you are the driver of that vehicle. It may move out unexpectedly and if it does you probably will never move again.

NOTE.—For additional information on the subject of this chapter see TF 25–394.

INDEX

INDEX

INDEX

Made in the USA
Monee, IL
01 November 2026

67205940R00094